Medical Physics

The cover photo shows a coloured three-dimensional MRI image, built up from successive 'slices' (two-dimensional scans) of a human head.

Medical Physics

Mike Crundell
Kevin Proctor

JOHN
MURRAY

Other titles in this series:
Astrophysics 0 7195 8590 2
Particle Physics 0 7195 8589 9

First published by
John Murray Publishers, a member of the Hodder Headline Group
338 Euston Road
London NW1 3BH

Layouts by Wearset Ltd, Boldon, Tyne and Wear
Illustrations by Bede Illustration
Cover design by John Townson/Creation

Typeset in 11.5/13pt Goudy by Wearset Ltd, Boldon, Tyne and Wear
Printed and bound in Spain by Bookprint, S.L., Barcelona

A catalogue entry for this title is available from the British Library

ISBN 0 7195 8591 0

Contents

Introduction

This book has been written principally for A-level students who are studying either Health Physics or Medical Physics as part of their Physics syllabus. It is also appropriate for students taking International examinations at a level equivalent to Advanced level.

The book does not attempt to approach the content from first principles. Rather, it assumes a knowledge and understanding of 'core physics' and the mathematical skills required for this 'core'. In particular, students should be familiar with elements of mechanics, waves and vibrations, radioactivity, and atomic and nuclear physics. Use of the exponential function is required. A number of worked examples are provided throughout the text, and at the end of each chapter there are self-assessment questions that allow students to test their understanding. These questions have been written so that they are at a standard appropriate to Advanced level.

The contents of the book provide more material than is likely to be required for any one particular Advanced level syllabus. On the other hand, a book of this size cannot possibly give comprehensive coverage of such a broad subject and is, necessarily, selective in content.

It is hoped that *Medical Physics* will not only provide readers with the necessary information and understanding to successfully complete their courses of study, but will also inspire them to want to study further the science of medicine.

Mike Crundell
Kevin Proctor

Acknowledgements

We are indebted to the many people who have been engaged in the production of *Medical Physics* as well as those who gave support and encouragement during the writing of the manuscript. We would particularly like to thank Katie Mackenzie Stuart (Science Publisher) for initiating the project and for her guidance, and also Jane Roth for her skilful editing that has done so much to improve the appearance of the book.

Finally, we would like to thank Leigh who had to endure hours of discussions rather than be entertained with lively conversation.

Mike Crundell
Kevin Proctor

Photo credits

Thanks are due to the following for permission to reproduce copyright photographs.

Cover CNRI/Science Photo Library; **p.1** *tl* and *bl* Science Photo Library, *r* Corbis UK Ltd; **p.12** Peter Gould; **p.23** Jeff Moore (jeff@jmal.co.uk); **p.24** Science Photo Library; **p.26** Bubbles; **p.42, p.49** *both* and **p.52** Science Photo Library; **p.55** PA News Photos; **p.60** Bridgewater Books (Dortenzio@bridgewaterbooks.co.uk); **p.68** Science Photo Library; **p.82** The Royal Surrey County Hospital NHS Trust; **p.84, p.90, p.92, p.94, p.102, p.103, p.105, p.106, p.109, p.112, p.119, p.124** *all*, **p.125, p.128** and **p.130** Science Photo Library; **p.131** The Royal Surrey County Hospital NHS Trust; **p.140** Science Photo Library, **p.142** John Townson/Creation, **p.144** Wellcome Trust, **p.150** and **p.152** Science Photo Library.

t = top, *b* = bottom, *l* = left, *r* = right.

Every effort has been made to contact copyright holders. The publishers apologise for any omissions which they will be pleased to rectify at the earliest opportunity.

Picture researcher: Liz Moore (lm@macunlimited.net)

The eye and sight

In this chapter you will read about:

- the basic structure of the eye
- how the eye forms a focused image
- the response of the eye to different wavelengths and intensities of light
- defects of the eye and associated corrective treatment

Introduction

In order to understand how the eye works it is first necessary to revise some of the properties of light. Light is a member of the electromagnetic spectrum and so has a wavelike nature. It may be reflected, refracted and diffracted, and it travels in straight lines through a medium of uniform optical density (that is, one through which the speed of light is constant). Refraction is the property of light that is responsible for the focusing processes that occur in the eye. When light travels from one medium into a medium of greater or lesser optical density, refraction occurs as light travels at a different speed in the second medium. Unless light is incident at an angle of 90° to the plane of the boundary between the two different media, it will undergo a change in direction at that boundary. The eye has parts with different optical densities and so, together with the curvature of its front surface, is able to perform all of the refraction necessary for the formation of an image.

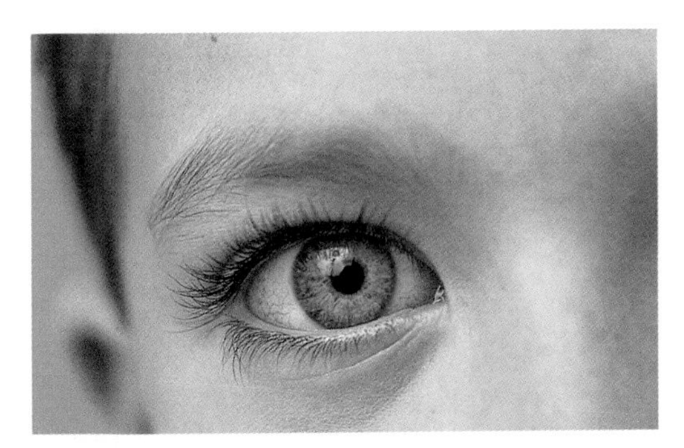

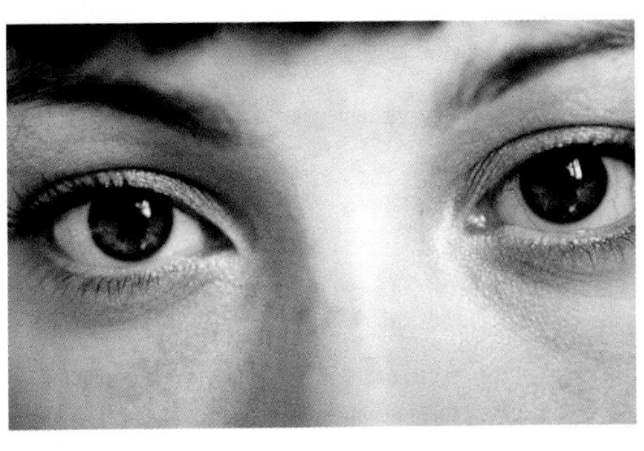

Figure 1.1

Where the eye requires additional help in forming a clear image, lenses (usually made from glass or plastic) are employed. Glass and plastic have different optical densities from air and so there is refraction of light when it is incident upon an air–glass or air–plastic boundary at angles other than 90°. The amount of refraction also depends upon the curvature of the lens. A lens that causes a greater change in direction of the incident ray of light has a greater curvature and hence a greater **power**. Conversely, a thinner, less curved lens has a lower power and so refracts the light less.

A lens that converges rays of light is called a **convex** lens (Figure 1.2a). The parallel lines in front of the lens represent the path followed by rays of light from a distant object. These lines are refracted at the lens surfaces due to the change in speed of the light as it enters and leaves the lens and due to the curvature at the faces of the lens. The rays converge and continue on through the point F.

A lens that diverges rays of light is called a **concave** lens (Figure 1.2b). The rays spread out after passing through the lens and appear to come from a point F. The broken lines denote the **apparent** path travelled by the diverging rays and *not* the actual path.

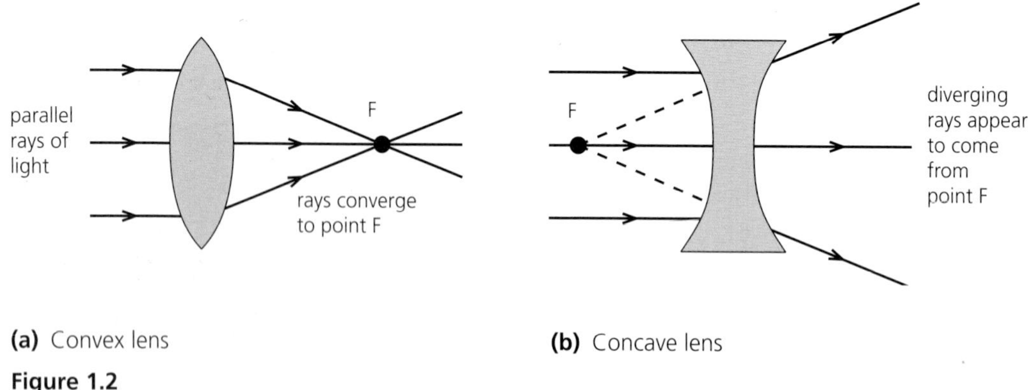

(a) Convex lens

(b) Concave lens

Figure 1.2

The formation of a real image

A **real** image is one that is formed when divergent rays of light, reflected (or emitted) from the same point on an object, are converged in such a way that they meet once again at a point. This convergence may be achieved by the use of a convex lens (see Figure 1.4) and the real image produced may be focused onto a screen.

If the rays of light from an object *appear* to originate from a point, as in the case of the image formed by a plane mirror (Figure 1.3), the image is said to be **virtual** or **not real**.

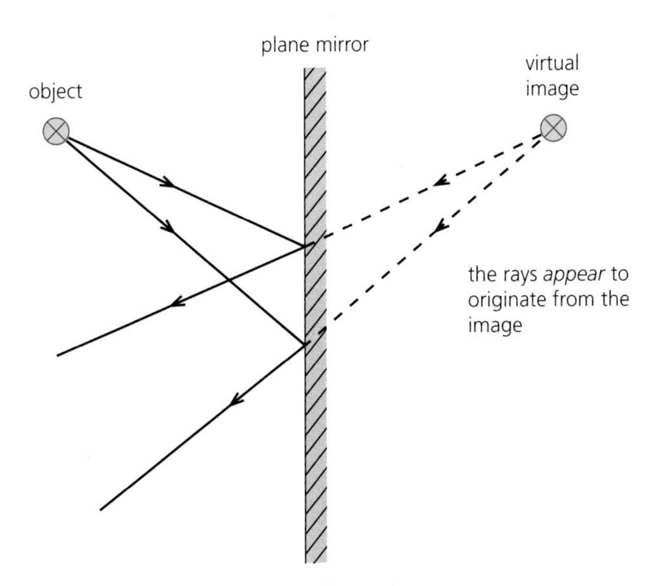

Figure 1.3 The formation of a virtual image by a plane mirror

The location of a real image formed by a convex lens

Refraction occurs at the front surface of the lens as the light passes from a medium of low optical density (in this case air) into a medium of high optical density (glass). Refraction also occurs at the second surface of the lens as the light leaves the glass and travels into the air. The amount of refraction that occurs depends on the difference in optical densities between air and glass, the curvature of the lens and the angle at which the incident ray hits the lens. The reference line in Figure 1.4 normal to the plane of the lens and passing through the optical centre of the lens is known as the **principal axis** of the lens. Rays of light that are parallel and close to the principal axis are refracted such that they all pass through a point on this axis called the **principal focus** or **focal point** (F). The distance from the principal focus to the centre of the lens is called the **focal length** (f) of the lens.

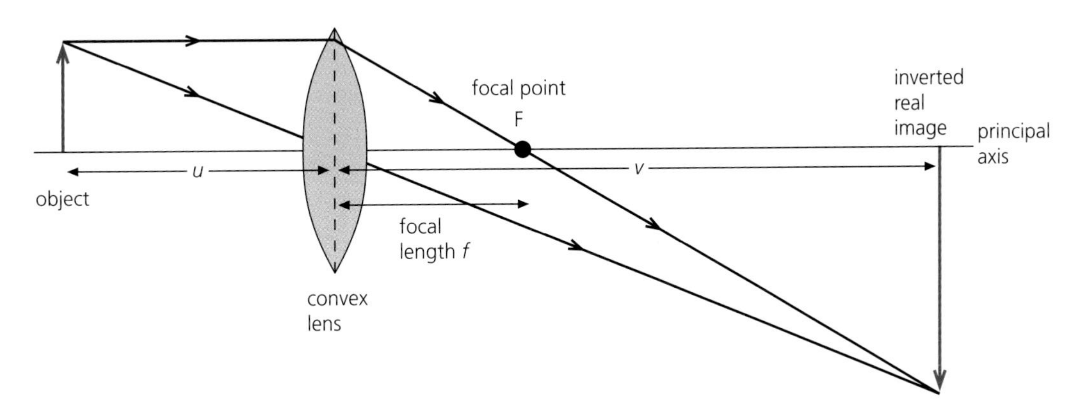

Figure 1.4 The formation of a real image by a convex lens

The distance u of the object from the centre of the lens is related to the distance v of the image from the centre of the lens by the formula

$$\frac{1}{u}+\frac{1}{v}=\frac{1}{f}$$

where f is the focal length of the lens.

When an image is formed of an object at infinity, the image position is at the focal point (Figure 1.5a). For all other object distances, the distance of the image from the lens is greater than the focal length (Figure 1.5b).

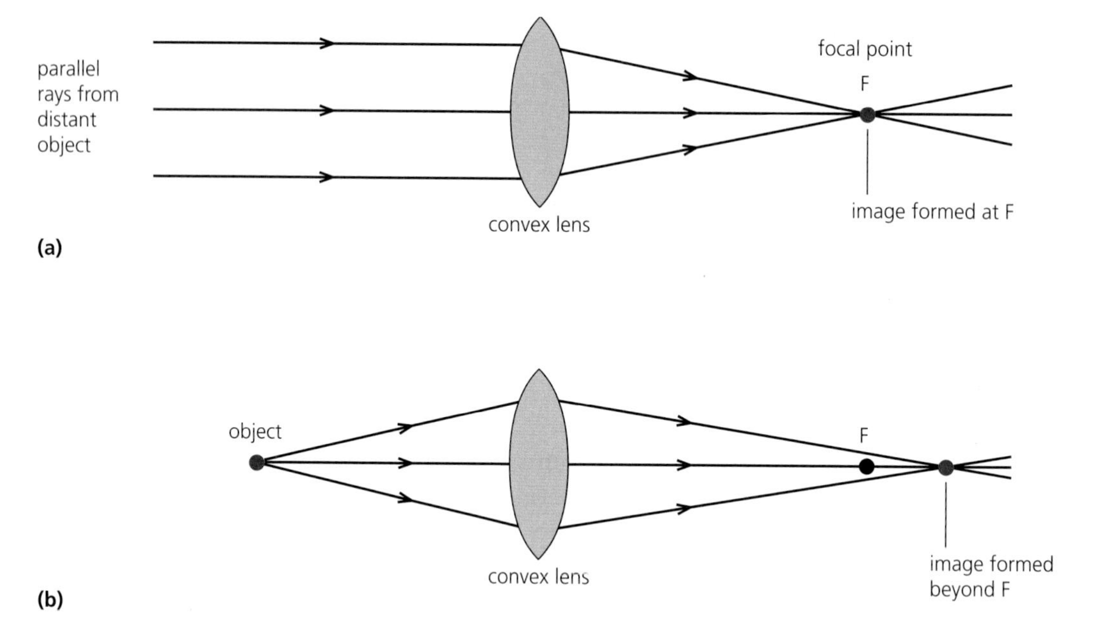

Figure 1.5

The power of a lens

For the specific case of the eye it is necessary to maintain a constant lens-to-image distance v, as the shape of the eyeball cannot change and the lens cannot move. As an object approaches the lens, therefore, it is necessary to increase the amount of refraction occurring at the lens or, in other words, to increase the **power** of the lens. The eye is able to change the power of its lens to enable it to maintain a focused image of a continually moving object. This process is called **accommodation** and is discussed later in this chapter (page 9). The power P of a lens is related to the focal length f by:

$$P=\frac{1}{f}$$

WORKED EXAMPLE 1.1

a An object is situated 180 cm in front of a convex lens of focal length 2.0 cm. Calculate the distance from the lens at which the image is formed.

b The object is subsequently moved to a position 12 cm in front of the convex lens. Find the new image position.

a $u = 180\,\text{cm}, \quad f = 2.0\,\text{cm}$

$$\frac{1}{u} + \frac{1}{v} = \frac{1}{f}$$

Rearranging the equation gives

$$\frac{1}{v} = \frac{1}{f} - \frac{1}{u}$$

Substituting values into the equation,

$$\frac{1}{v} = \frac{1}{2.0} - \frac{1}{180}$$

so

$$\frac{1}{v} = 0.49\,\text{cm}^{-1}$$

and

$$v = 2.02\,\text{cm}$$

b $u = 12\,\text{cm}, \quad f = 2.0\,\text{cm}$

$$\frac{1}{v} = \frac{1}{f} - \frac{1}{u}$$

Substituting values into the equation,

$$\frac{1}{v} = \frac{1}{2.0} - \frac{1}{12}$$

so

$$\frac{1}{v} = 0.42\,\text{cm}^{-1}$$

and

$$v = 2.4\,\text{cm}$$

As an object approaches a convex lens, its image position moves further from the lens.

In the equation $P = 1/f$, if the focal length is measured in metres then the unit of power is the **dioptre** (D). A lens with a large convex curvature on both faces will have a greater power than a lens of the same material with a smaller convex curvature on both faces.

The focal length f and the image distance v for a concave lens both have negative values (Figure 1.6). As the power of a lens is calculated from the reciprocal of its focal length in metres, so the power of a concave lens has a negative value. A lens of large positive power (+40 D) converges light by a large amount compared with a lens of low positive power (+5 D). A lens of large negative power (−50 D) diverges light by a greater amount than a lens of low negative power (−5 D).

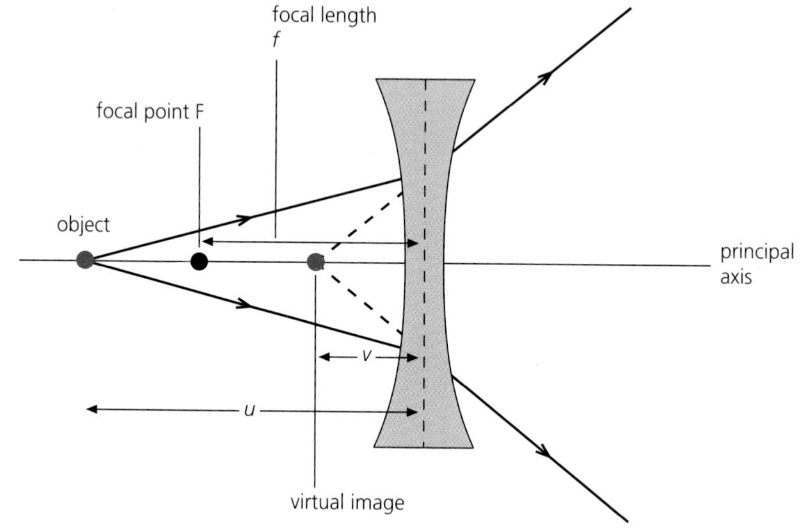

Figure 1.6 The formation of a virtual image by a concave lens

The structure of the eye

The structure of the eye is similar to that of a simple camera (Figure 1.7). Light enters the eye through an opening called the pupil in the same way that light enters the camera through the aperture. The pupil diameter is involuntarily controlled, increasing in low-intensity lighting and decreasing in high-intensity lighting. The refracting system of the eye acts in the same way as a converging lens (see Figure 1.4) and an inverted image is formed at the back of the eye.

Figure 1.8 shows the eye in more detail. The **cornea** (front surface of the eye) is curved and it is here that most refraction occurs. Light passes through the aqueous humour (a transparent liquid), through the pupil (the opening in the iris) and is further refracted by the lens before passing into a dark chamber filled with another transparent fluid called vitreous humour. The fluids in the eye provide nutrients and also help the eye to maintain its shape. The image is finally formed on a light-sensitive surface at the back of the eye, called the **retina**. This surface contains millions of nerve endings that stem from two types of cell known as rods and cones.

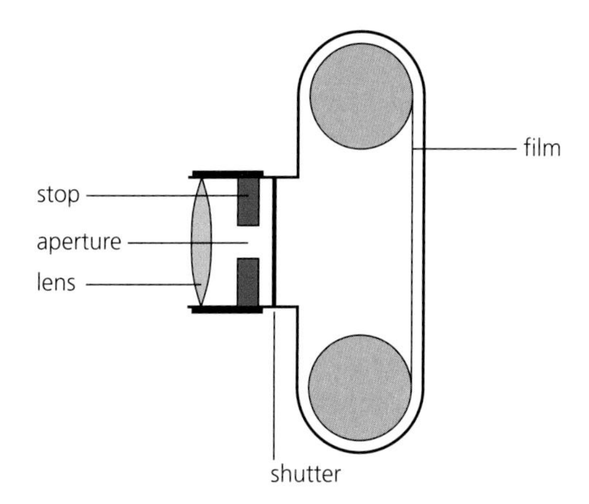

Figure 1.7 A simple camera

The nerve fibres in the retina leave the eye at a point on the retina called the **blind spot** (so called because there are no nerve endings at this point and as a consequence no light is detected here). This bundle of fibres, called the **optic nerve**, carries information to the brain. The brain inverts the image and allows an interpretation that is dependent on the extent to which each rod or cone has been stimulated.

When the eye focuses on an object, the image is usually formed on the most sensitive part of the retina called the **fovea**. This is a small depression in the retina that contains only cones, and it is here that the greatest detail is provided when an image is formed on it.

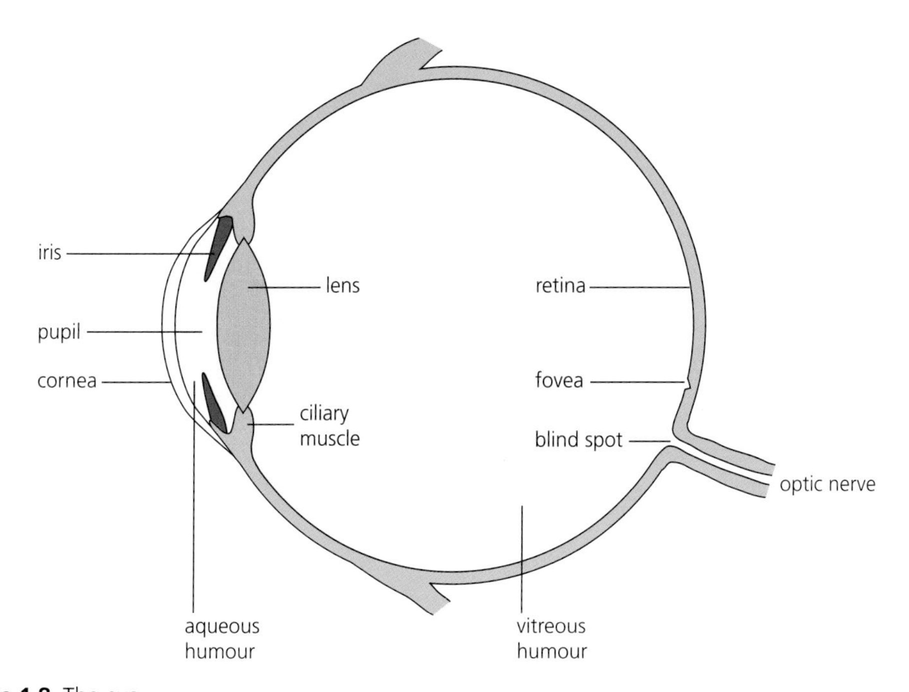

Figure 1.8 The eye

WORKED EXAMPLE 1.2

The refracting system of the eye causes a clear image of an object at infinity to be formed at the retina. The distance of the retina from the cornea is 1.9 cm. Assume that the eye acts as a single thin lens, situated at the front surface of the cornea (see Figure 1.9).

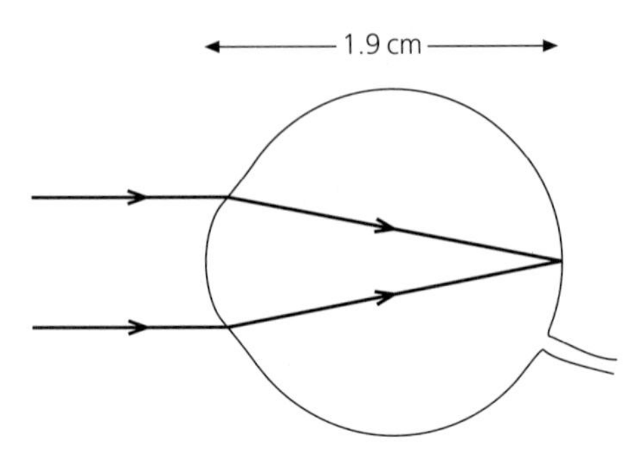

Figure 1.9

a Calculate the focal length of the refracting system of this eye.
b Calculate the power of this eye when viewing an object at infinity.

a $u = \infty, \quad v = 0.019\,\text{m}$

$$\frac{1}{f} = \frac{1}{u} + \frac{1}{v}$$

$$\frac{1}{f} = \frac{1}{\infty} + \frac{1}{0.019}$$

$$\frac{1}{f} = \frac{1}{0.019} \quad \text{since } \frac{1}{\infty} = 0$$

so

$$f = 0.019\,\text{m}$$

b

$$P = \frac{1}{f}$$

$$= \frac{1}{0.019}$$

so

$$P = 52.6\,\text{D}$$

Remember that the focal length f must be measured in metres to give the power P in dioptres.

Accommodation of the eye

The eye has a structure that enables it to vary its focal length and hence to focus on an object first at infinity and then at a point close to the eye. This ability to vary the focal length is called **accommodation**. The eye is said to be **accommodated** when focusing on an object at the **near point**. The near point is the closest position to the eye that an object may be placed, in order for that eye to focus comfortably on the object. For an average person, the near point is about 25 cm from the eye. The eye is **unaccommodated** when viewing an object at the **far point**. The far point is the position furthest from the eye that an object may be viewed clearly. For the average normal eye the far point is at infinity (∞).

Most refraction occurs at the front surface of the eye where the air is in contact with the cornea. There is a large difference in the refractive indices of air and the cornea and so a large amount of refraction occurs at this interface. The refraction at this boundary is responsible for about +46 D of the power of the refracting system of the eye. At the other interfaces within the eye refraction occurs to a lesser degree. Curvature of the eye lens is responsible for a small amount of refraction. This small amount of refraction may be varied slightly by changing the curvature of the front and rear surfaces of the lens. The **ciliary muscles** around the lens (see Figure 1.8) are responsible for this change in shape of the lens. These muscles relax to allow the lens to flatten and the eye to focus on an object at infinity (Figure 1.10a). When the eye is viewing an object that is near, the ciliary muscles contract to cause the lens to bulge (Figure 1.10b).

The eye of a middle-aged adult has its lowest power of about +59 D when viewing an object at infinity and it may change its power by about +4 D to +63 D in order to view an object at the near point.

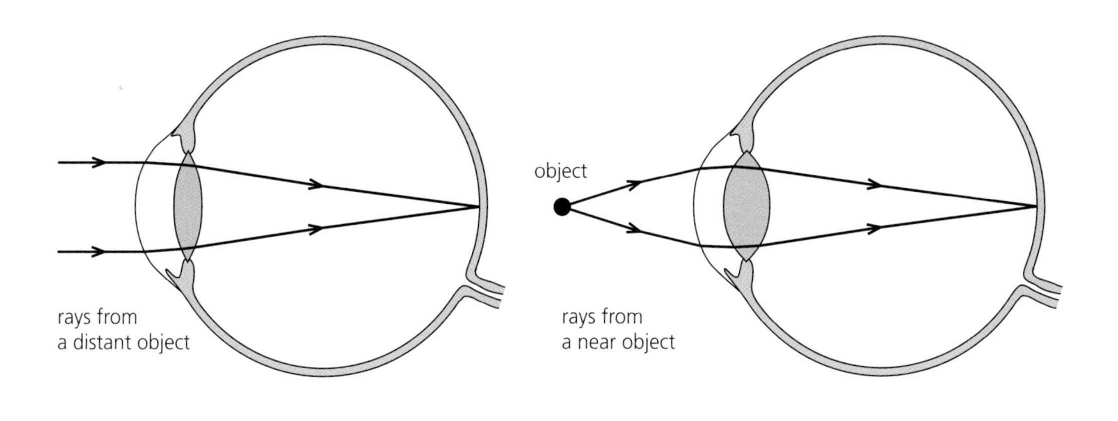

(a) Unaccommodated eye

(b) Accommodated eye

Figure 1.10

Depth of field and depth of focus

When the eye is focusing on an object that is moving towards it, the eye lens must keep increasing its power in order for a sharp image to remain on the retina. This is an involuntary procedure for most people. At certain points during the movement of an object towards (or away from) the eye, the image on the retina will become unacceptable and adjustments to the power of the eye lens will be made to bring the image into focus.

When the distance of the object from the eye is large, however, a small change in this distance does not necessitate a change in power of the eye lens. This is because the image formed on the retina is still acceptably clear. The **depth of field** is the range of object distances for which an acceptably clear image is formed on the retina *without* a change in power of the eye lens (Figure 1.11).

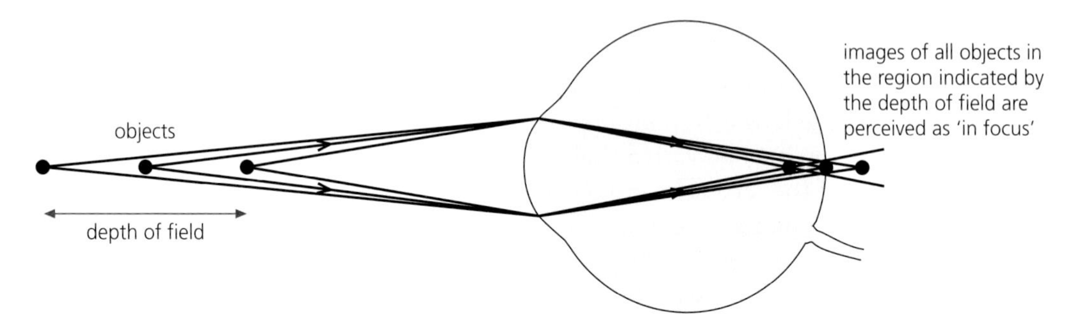

Figure 1.11 Depth of field

When the position of a focused image is just in front of, or just behind, the retina, the image at the retina will be slightly out of focus (or blurred). The eye lens accommodates to decrease or increase the power of the refracting system of the eye in order to site the image on the retina. However, it is possible for the eye to perceive an image formed just in front of (or just behind) the retina as acceptably clear. The range of image distances for a given object in which an image may be formed and perceived as acceptably clear by the eye without a change in power is called the **depth of focus** (Figure 1.12).

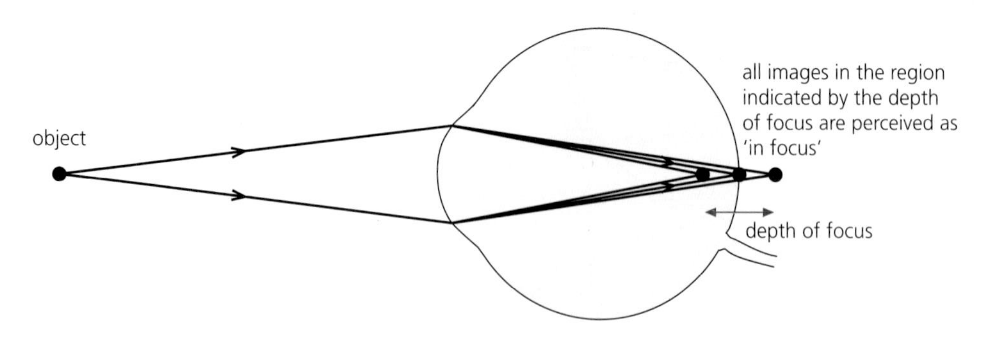

Figure 1.12 Depth of focus

Resolution of the eye

When a person views a very distant car at night, the light from the two headlights of the car appears to come from a single point source. As the car gets nearer, the person is able to distinguish the two sources of light as separate and is said to be able to **resolve** the two sources of light. The **resolution** of the eye is the angle subtended at the eye due to light from two point objects that the eye can just resolve.

WORKED EXAMPLE 1.3

An eye can just detect two objects that are 2.0 mm apart as being separate when they are at a distance of 10 m (Figure 1.13). Calculate the resolution of the eye in radians.

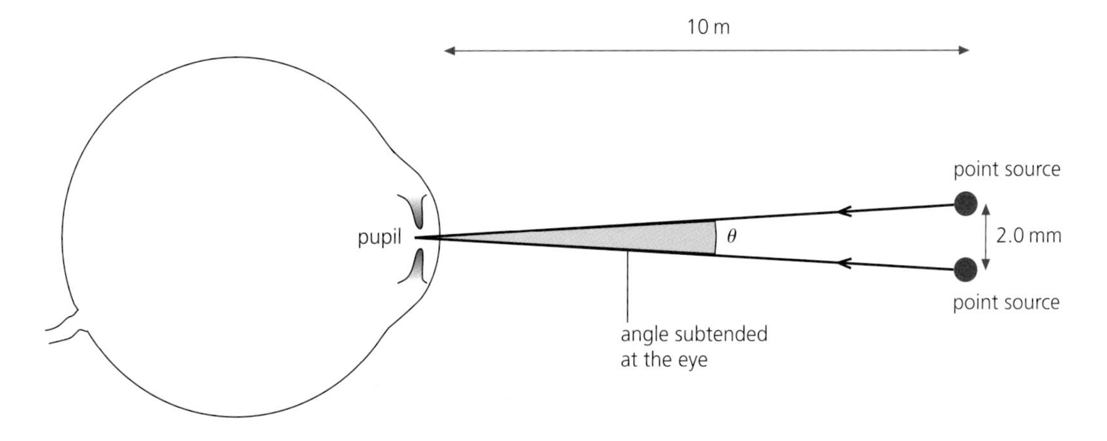

Figure 1.13

The resolution in radians is the arc divided by the radius:

$$\text{resolution} = \frac{2.0 \times 10^{-3}}{10} = 2.0 \times 10^{-4} \text{ radians}$$

The image formed at the retina is not a straightforward image. It is complicated by diffraction at the pupil. The resulting diffraction pattern due to a point source of light is shown in Figure 1.14a overleaf. Figure 1.14b shows how the intensity of light varies with distance along the image.

It is helpful to revisit the theory behind diffraction of monochromatic light at a slit, in order to understand the dependence of resolution on diffraction. See Box 1.1.

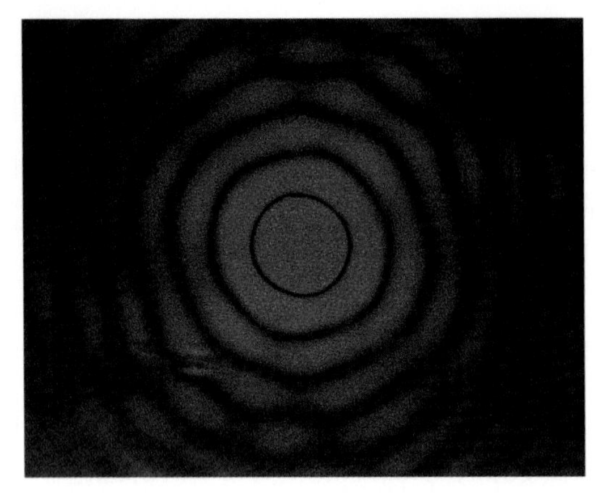

(a) The image of a point source is a diffraction pattern

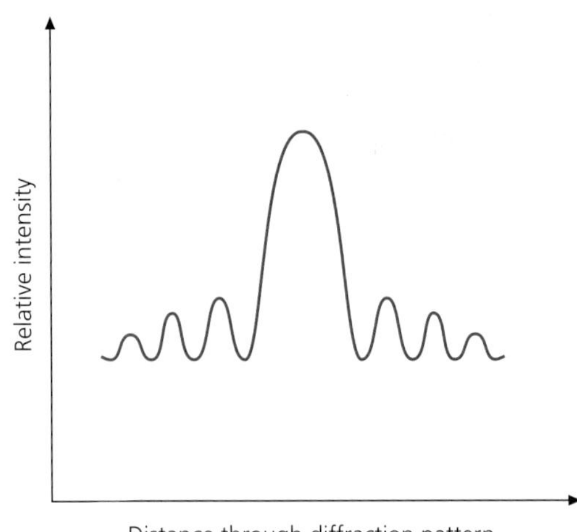

Figure 1.14

(b) The variation of intensity across the image

Box 1.1 Diffraction of monochromatic light at a slit

Figure 1.15 shows a monochromatic (single wavelength) light source, slit and screen with the maximum (B) and minimum (D) intensities of the diffraction pattern marked.

Figure 1.15 Diffraction at a slit

The light that is diffracted at the slit in the direction of the first minimum undergoes destructive interference such that the resulting intensity at the diffraction pattern is zero. The light leaving the slit in the direction of this first minimum may be considered as many parallel rays. If each ray can be paired with another ray from the slit of path difference $\lambda/2$, then all of the rays arriving at the position of the first minimum undergo destructive interference. Figure 1.16a shows light emerging from the slit in the direction of the first minimum D, along a line that makes an angle θ with the normal to the slit. Figure 1.16b shows how two rays of light from the slit may be paired together. The extra path travelled by ray ED is $EF = \lambda/2$. This means that when the two rays meet at D, they will be 180° out of phase, causing destructive interference and zero intensity at that point. Similar pairs of rays separated by a distance of $d/2$, where d is the slit width, for example from points C and G (or K and H, see Figure 1.16c), along the length of the slit aperture will also cause destructive interference.

The wavelength of the light, λ, slit width d and angle θ (subtended at the slit between the path of the central bright maximum and the first minimum) are related by the equation:

$$\sin\theta = \frac{\lambda}{d}$$

which is derived from the triangle AEF in Figure 1.16b. For very small angles θ, $\sin\theta$ is approximately equal to θ. The equation then becomes:

$$\theta = \frac{\lambda}{d}$$

For a circular aperture such as the eye's pupil,

$$\theta = 1.22 \times \frac{\lambda}{d}$$

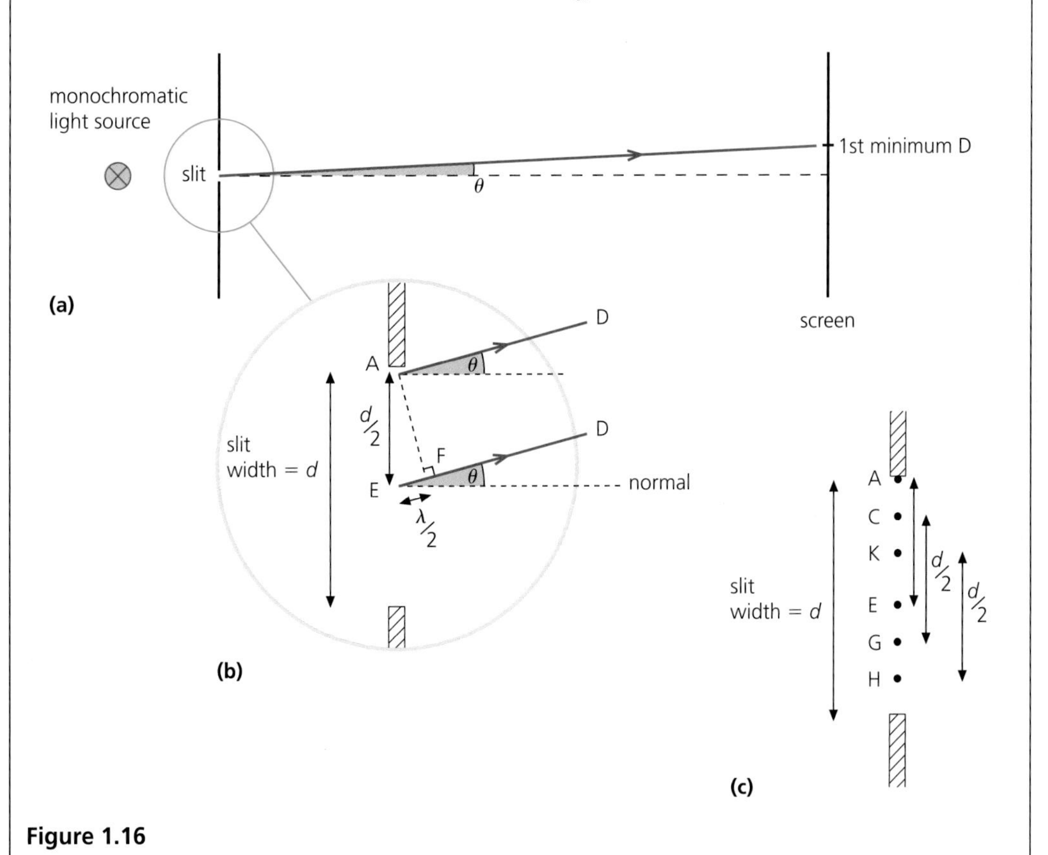

Figure 1.16

Two point sources of light that are close together and situated a long way away from the eye may be perceived as a single source of light by the eye. An example of this occurs when viewing a double star that appears to be a single spot of light. The condition that needs to be met if the two sources are to be resolved as being separate is called the **Rayleigh Criterion**. The Rayleigh Criterion suggests that the two sources are perceived to be resolved if the maximum intensity due to one diffraction pattern is positioned over the first minimum due to the diffraction pattern of the second source (Figure 1.17a). If the light from the two sources has the same wavelength λ, the angle θ subtended at the eye, of pupil diameter d, by light travelling along the path to the central maximum and the path to the first minimum is given by $\theta = 1.22\lambda/d$ (see Box 1.1). Thus, for resolution of the two sources according to the Rayleigh Criterion, their angular separation must be greater than or equal to this value.

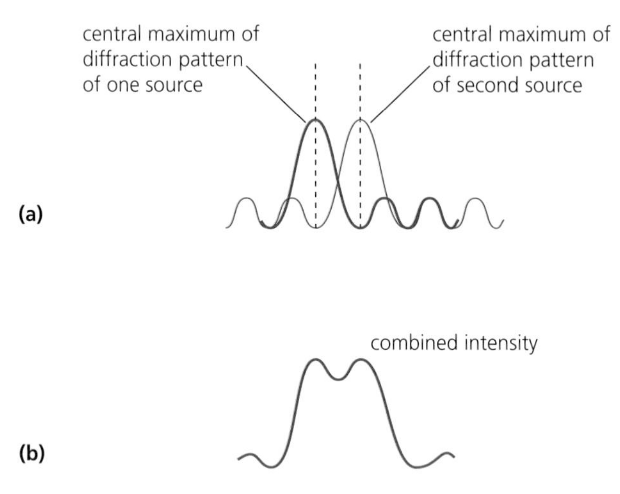

Figure 1.17 The Rayleigh Criterion for the resolution of two point sources

The combined intensity due to the two 'just resolved' diffraction patterns is shown in Figure 1.17b. If the minimum between the two peaks in this combined diffraction pattern is to be detected on the retina (at point Q in Figure 1.18), then light from this minimum must fall on a cone with light from the peaks either side (points P and R) falling on neighbouring cones. This means that the ultimate resolution of the eye in practice is determined by the separation of the cones in the fovea, which is a distance of about 2.5×10^{-6} m. The angular separation of the two sources when they are just resolved by the retina is therefore

$$\theta = \frac{\text{PR}}{\text{eye diameter}} = \frac{2 \times \text{cone spacing}}{\text{eye diameter}}$$

For an eye of diameter 1.9 cm, the resolution will therefore be $2 \times 2.5 \times 10^{-6}/1.9 \times 10^{-2}$ which is approximately 2.5×10^{-4} radians.

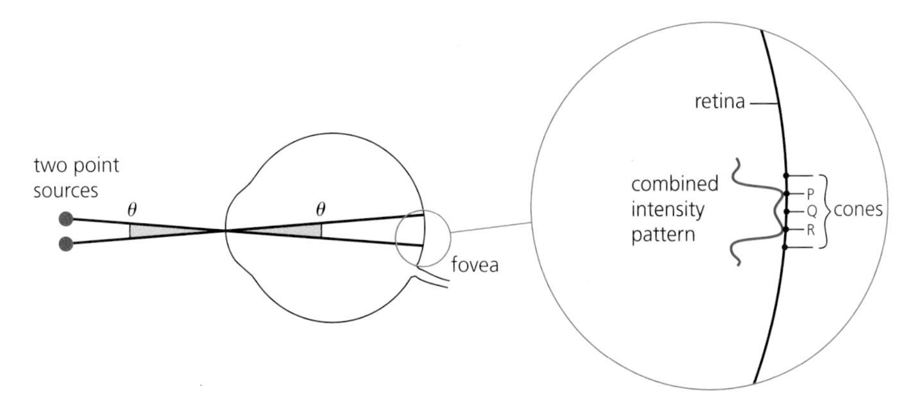

Figure 1.18 The resolution of the eye is dependent on the cone spacing

Response of the eye to variations in wavelength and intensity

The rods and cones in the retina both respond in degrees to variation in light intensity. Light-sensitive chemicals in the rods and cones decompose and cause electrical impulses that are carried along the respective nerve fibres to the brain.

Rods do not differentiate between colours. Their response across the visible spectrum is, however, not linear. They are most sensitive in the green region where the fraction of incident photons absorbed is the greatest (see Figure 1.19). In other regions of the visible spectrum, the fraction of incident photons absorbed is less. The chance of an individual photon being detected increases as the wavelength of the photon approaches 510 nm and hence the response of the eye due to the rods is most sensitive at about 510 nm.

Rods have a greater sensitivity in low light conditions compared with cones (see Figure 1.20). In low light intensities, the rods are responsible for most of the vision. Vision at very low light intensities is called **scotopic vision**. There are no rods on the fovea. As the distance from the fovea across the retina's surface increases, the proportion of rods to cones increases. The consequence of this is that in very low-intensity light conditions, there is no vision directly in front of the eye and only vague vision around this region.

Cones require a higher intensity of light in order to operate efficiently. The fovea contains only cones and it is the stimulation of these cells that is responsible for detail in an image. There are three types of cone, each having a maximum sensitivity in a different region of the visible spectrum (Figure 1.21).

The peaks of these three types of cone lie in the red, green and blue regions of the spectrum and so the cones are known respectively as **red**, **green** and **blue cones**. Cones are responsible for colour discrimination or **photopic vision**. The interpretation of the colour of light falling on an area of the retina is dependent on the fraction of the absorption of this light by each type of cone. The adding of information from each type of cone leads to the perception of a colour with a dominant wavelength in much the same way as differing intensities of the red, green and blue dots on a television screen add together to give the impression of a colour.

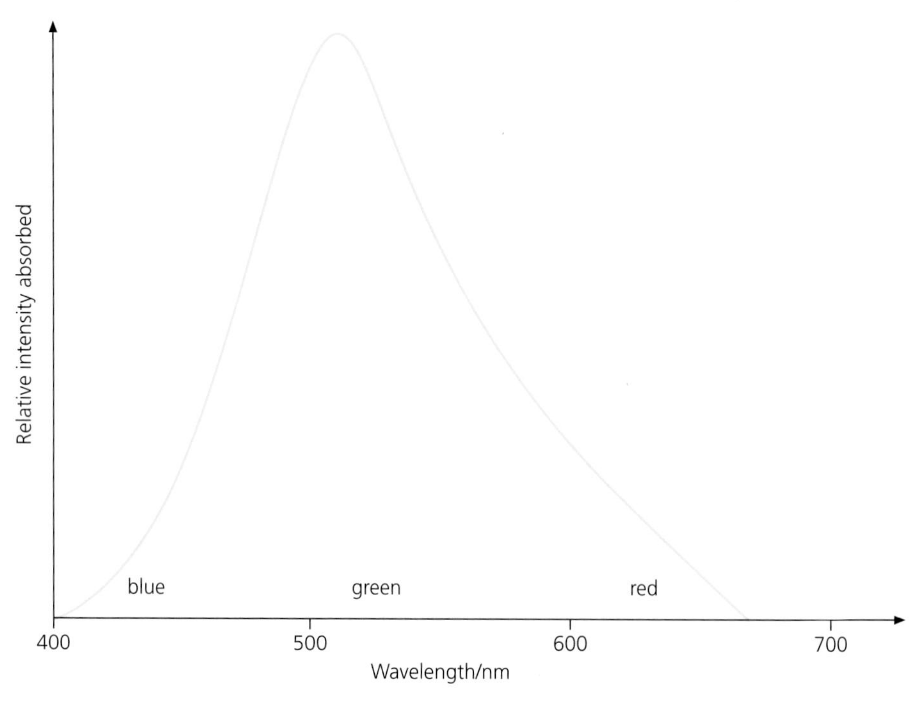

Figure 1.19 Sensitivity of rods to the visible wavelength range (in terms of proportion of light absorbed)

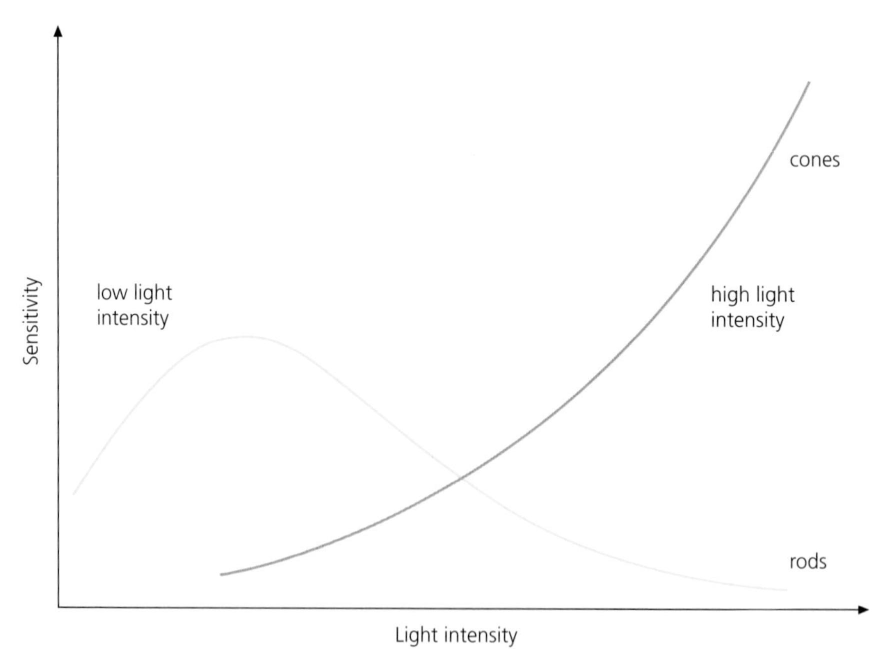

Figure 1.20 Sensitivity of rods and cones in different light intensities

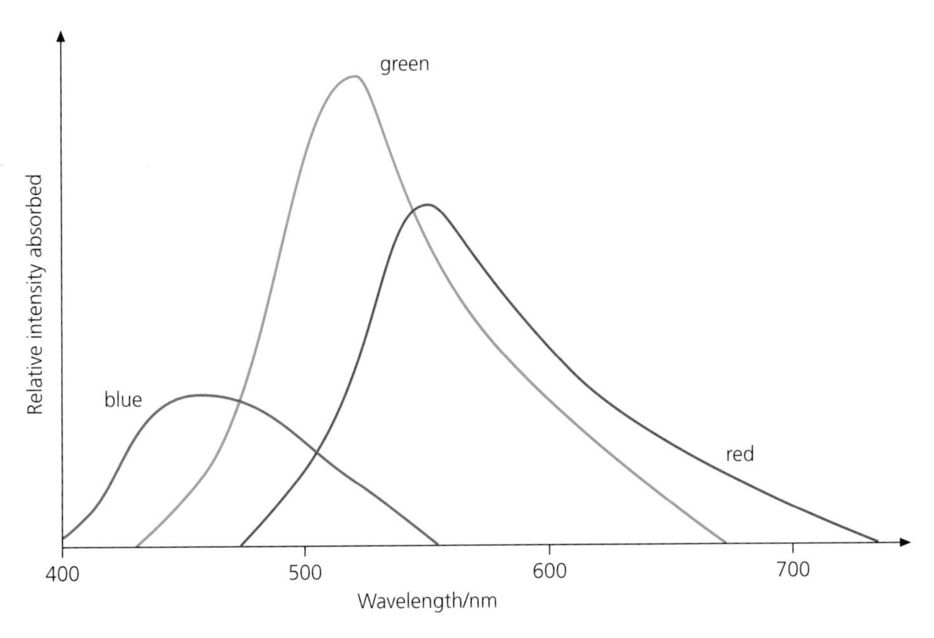

Figure 1.21 Sensitivity of the three types of cone to visible wavelengths (in terms of proportion of light absorbed)

Defects of the eye

There are a number of defects that affect normal vision. Some of these may be easily rectified with the use of spectacle lenses, while others have no cure. Recent developments in laser surgery have provided a permanent remedy for some defects (see Figure 10.5, page 144). The five most common eye defects are explained below.

Short sight (myopia)

Many eye problems result in the image of an object that is being viewed appearing blurred to the viewer. For a **short-sighted** person, objects that are close to the eye are in focus but blurring occurs when trying to focus on a distant object. The image of the distant object is formed in front of the retina for a short-sighted person (Figure 1.22a). Rays of light continue on through the image and then diverge. When light eventually lands on the retina, the image is larger and less well defined, in other words, it is out of focus. To correct the defect, it is necessary to diverge the light prior to entry into the eye, resulting in a lower overall power of the refracting system of the eye (Figure 1.22b). A diverging or concave lens with a negative power is employed to correct short sight.

A corrective lens is chosen such that when its power is added to the power of the eye, the combined power produces an image on the retina. This may be summarised as an equation:

$$(\text{power of eye}) + (\text{power of corrective lens}) = (\text{total power of refracting system})$$

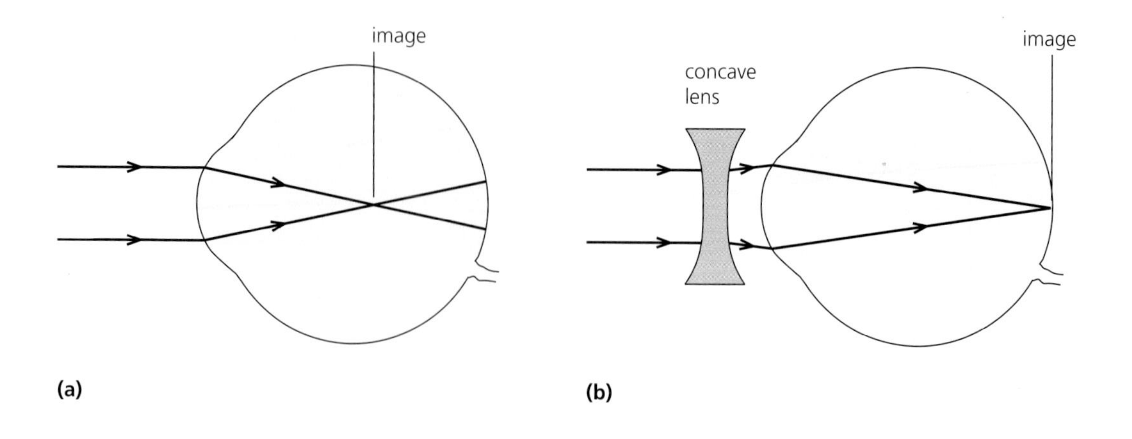

Figure 1.22 Short-sightedness and its correction with a diverging lens

WORKED EXAMPLE 1.4

An eye has a refractive power of 64 D when viewing an object at infinity. In order to produce a focused image of the distant object, the power of the eye needs to be 59 D. Calculate the power and state the type of the corrective lens required by this eye.

$$\begin{matrix}\text{power of unaided eye} \\ \text{when viewing object}\end{matrix} + \begin{matrix}\text{corrective lens} \\ \text{power } P\end{matrix} = \begin{matrix}\text{power required by eye} \\ \text{to view object}\end{matrix}$$

$$64\,\text{D} + P = 59\,\text{D}$$

so

$$P = 59\,\text{D} - 64\,\text{D} = -5\,\text{D}$$

The power of the corrective lens required is -5 D. The negative power indicates that the lens is a concave lens.

WORKED EXAMPLE 1.5

Assume that the eye in this example acts as a single thin lens situated at the front surface of the cornea and that the cornea-to-retina distance is 0.019 m (Figure 1.23). When viewing an object at infinity, the eye of a short-sighted person forms a blurred image. The lowest power that this eye is able to maintain means that the furthest distance at which an object may be viewed clearly is 0.60 m (this is the far point of the eye).

Calculate:

a the focal length of this eye when viewing an object at the far point of 0.60 m,

b the power of the eye when viewing an object at its far point,

c the power needed by this eye to view an object clearly at infinity,

d the power of the corrective lens which, when added to the unaided eye, will allow the eye to view clearly an object at infinity.

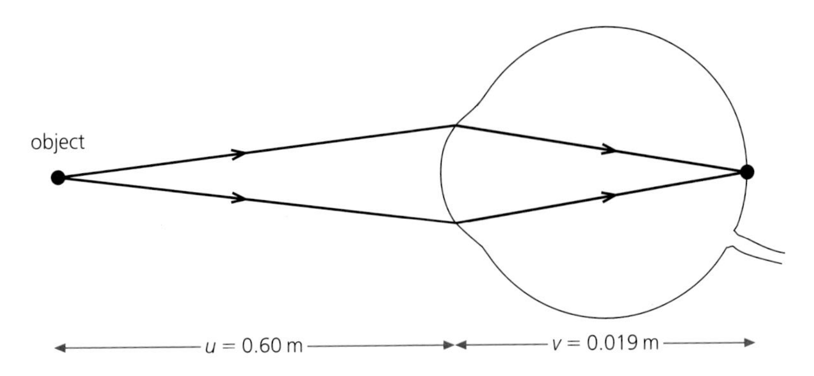

Figure 1.23

a

$$\frac{1}{f} = \frac{1}{u} + \frac{1}{v}$$

$$= \frac{1}{0.60} + \frac{1}{0.019}$$

$$= 54.3\,\text{m}^{-1}$$

so

$$f = 0.0184\,\text{m}$$

b

$$P = \frac{1}{f}$$

$$= \frac{1}{0.0184} = 54.3\,\text{D}$$

c The cornea-to-retina distance in the eye remains unchanged ($v = 0.019\,\text{cm}$).

$$P = \frac{1}{f} = \frac{1}{u} + \frac{1}{v}$$

$$= \frac{1}{\infty} + \frac{1}{0.019}$$

$$= \frac{1}{0.019} = 52.6\,\text{D}$$

d Power of unaided eye (when viewing at the far point) + power of corrective lens = required power to view object clearly at infinity. The eye needs a total power of 52.6 D. The unaided eye has a power of 54.3 D. It is therefore necessary to add a negative power lens to the eye in order to reduce the overall power of the eye.

$$52.6\,\text{D} - 54.3\,\text{D} = -1.7\,\text{D}$$

A lens of power -1.7 D is required.

Long sight (hypermetropia)

A **long-sighted** person is able to see clearly objects that are far away but objects that are close appear to be blurred. When the eye of a long-sighted person refracts rays of light from a near object, the image on the retina is not in focus. The position of the focused image formed by the eye is behind the retina (Figure 1.24a). In order to correct the defect, the light must be converged prior to entry into the eye (Figure 1.24b). A convex lens is chosen such that the overall power of the eye is increased. Once again it is necessary for the powers of the eye and corrective lens to be added together to allow the eye to focus clearly on a near object.

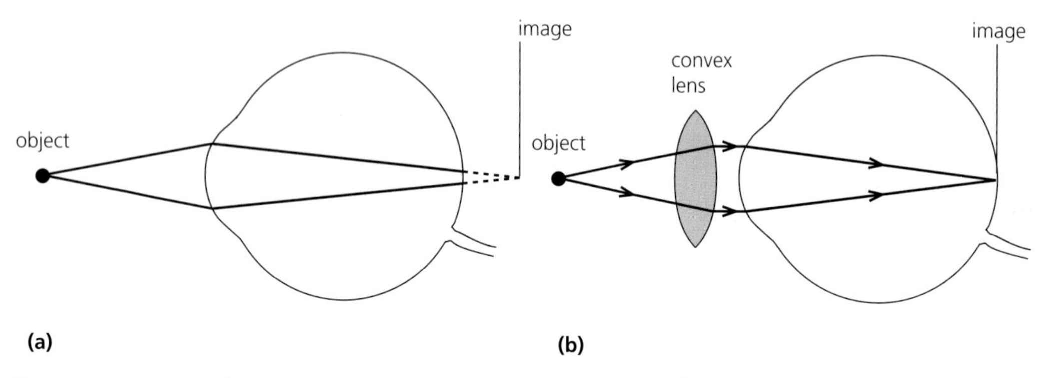

Figure 1.24 Long-sightedness and its correction with a converging lens

Astigmatism

The cornea has a surface that 'normally' has the same curvature in both the horizontal and vertical directions, refracting vertical and horizontal beams of light by the same amount (i.e. the refracting power in the horizontal plane is the same as that in the vertical). It is very common, however, to find that the cornea is more curved in one direction compared with another. This is called **astigmatism**.

Figure 1.25 shows light passing through an eye in which the curvature of the cornea in the z-direction is greater than the curvature in the y-direction. The power of the eye is thus greater in the z-direction than in the y-direction. Horizontal rays of light from an object such as a cross are refracted more (due to the larger curvature) and so form an image in front of the retina, while vertical rays are refracted normally to form an image on the retina. The effect on the sight of a person whose cornea has different powers in different planes as described above is that the vertical part of the cross will appear dark and sharp, while the horizontal part of the cross will appear lighter in colour and blurred.

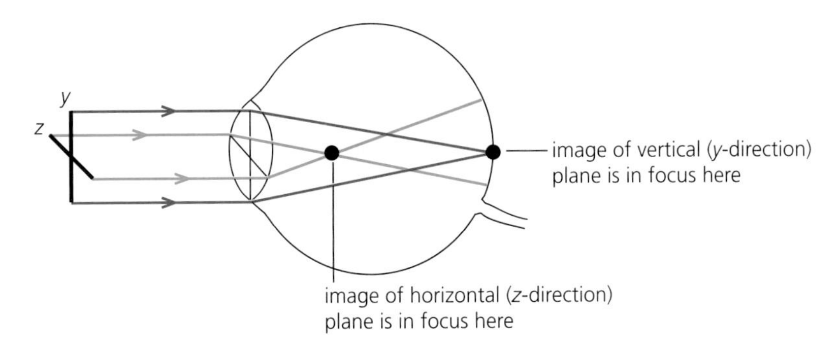

Figure 1.25 The effect of astigmatism

WORKED EXAMPLE 1.6

The eye of a long-sighted person has a near point of 60 cm. Calculate the power of the corrective lens needed to bring the near point to a distance of 25 cm from the eye. Assume that the refractive system of the eye acts as a single thin lens situated at the front of the cornea and that the distance from the cornea to the retina is 0.019 m.

The power of the unaided eye when viewing an object at the near point of 60 cm is given by:

$$P = \frac{1}{f} = \frac{1}{u} + \frac{1}{v}$$

$$= \frac{1}{0.60} + \frac{1}{0.019}$$

(Remember that u and v must be converted to a value measured in metres.)

$$P = 54.3\,\text{D}$$

The power of a normal eye when viewing an object at a normal near point of 25 cm is given by:

$$P = \frac{1}{f} = \frac{1}{u} + \frac{1}{v}$$

$$= \frac{1}{0.25} + \frac{1}{0.019}$$

$$= 56.6\,\text{D}$$

In order to bring the power of the unaided eye up to the power of a normal eye when viewing at the near point, it is necessary to increase the overall power by

$$56.6\,\text{D} - 54.3\,\text{D} = 2.3\,\text{D}$$

So a lens of power +2.3 D must be added to correct the defect of vision.

The corrective lens for an eye that suffers from astigmatism is cylindrical. Figure 1.26a shows the shape of the corrective cylindrical lens for the eye of Figure 1.25. The astigmatism of this eye causes the horizontal line of the cross to appear blurred when the vertical line is in focus. Light from the line in the horizontal plane must be diverged prior to entry into the eye. The corrective lens must be shaped so that the lens in this plane is concave. Light from the vertical plane must not be refracted and so the shape of the vertical section through the glass is rectangular. Figure 1.26b shows another type of cylindrical lens that compensates for a defect in the horizontal plane while allowing light from the vertical plane to pass through unaffected. With this lens the light from the horizontal plane is converged prior to entry into the eye.

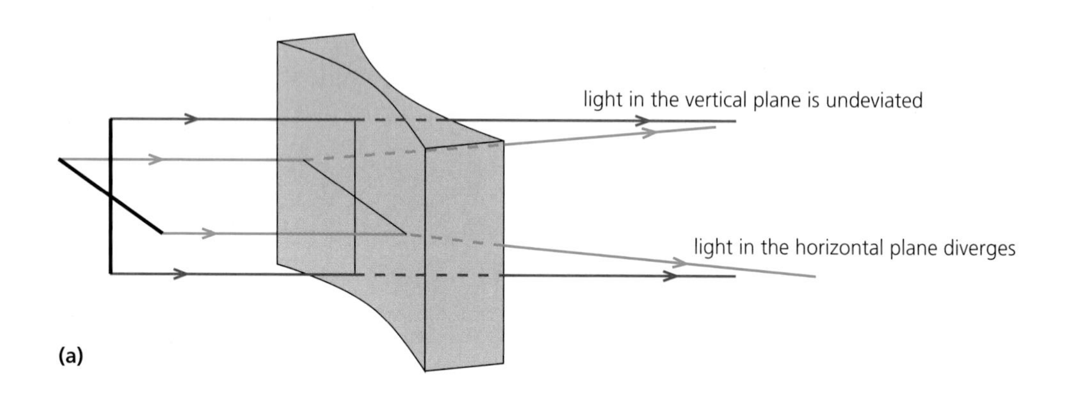

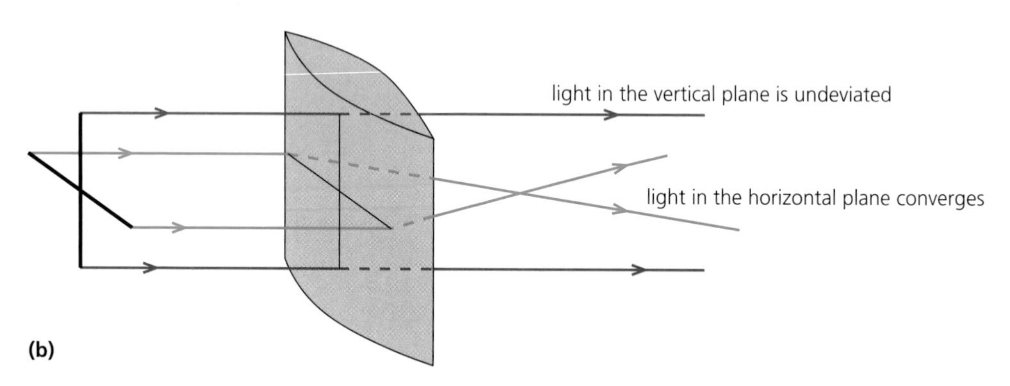

Figure 1.26 Corrective lenses for astigmatism

In order to detect the planes in which the eye's cornea is not spherical, a chart such as the one in Figure 1.27a is viewed. If the eye is short-sighted in the horizontal plane, the horizontal lines in the chart will appear blurred (Figure 1.27b). The vertical and near-vertical lines in the chart will appear darker and in focus.

(a) Chart for testing for astigmatism

(b) How the chart looks to someone with astigmatism in the horizontal plane

Figure 1.27

Presbyopia

Presbyopia is a defect of the eye associated with old age and is partly due to the hardening of the lens. This makes it less able to change shape and hence reduces the ability of the eye to accommodate. In a young eye, the lens can change the power of the optical system of the eye by as much as 14 D. This range of accommodation may reduce to zero in old age, causing the eye to experience problems in focusing on objects at infinity as well as at the near point.

Colour blindness

Colour blindness is a defect that affects about 1 in 12 males and 1 in 200 females. The degree of colour blindness varies from individual to individual, but essentially the problem lies in the brain's inability to distinguish between signals from one or two of the three cones. It is very rare to find a person who is unable to differentiate any colour, in other words to see in black and white (and shades of grey). The most common form of colour blindness involves the inability to distinguish between reds and greens. Figure 1.28 shows a test chart that is designed to identify confusion between red and green.

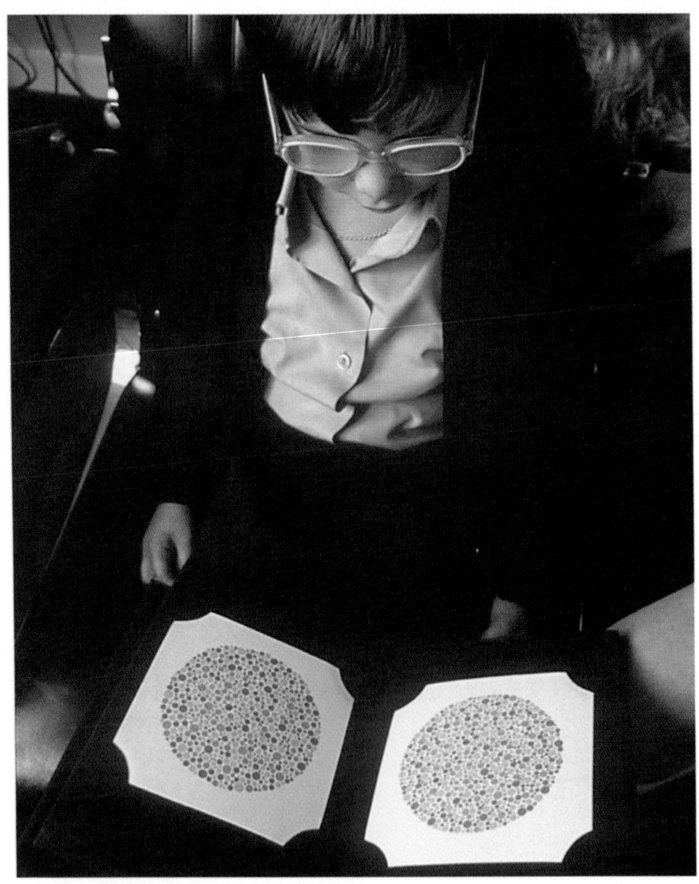

Figure 1.28 Testing for red–green colour blindness

Questions

1 a Draw a labelled diagram to show the main physical features of the human eye.
b By reference to your diagram, describe the processes that take place when
i) the eye focuses on an object at infinity, and
ii) the eye accommodates to focus on an object at the near point.

2 a Outline briefly how the eye is able to distinguish between light of different wavelengths.
b i) State the range of wavelengths detectable by the eye for a person with normal sight.
ii) State and explain how this range of wavelengths changes as the intensity of light falls to a very low value.

3 An eye has a near point of 15 cm and a far point of 50 cm. The distance of the retina from the front surface of the cornea is 0.019 m. For all calculations assume that the eye acts as a single thin lens situated at the front surface of the cornea.
a Calculate the power of the eye when focusing on an object at the far point of 50 cm.
b Calculate the power that this eye should have when focusing on an object at infinity.
c Calculate the power of the corrective lens needed by this eye to focus clearly on an object at infinity.
d State the eye defect and describe the shape of the corrective lens.
e Calculate the power of the unaided eye when viewing an object at its near point.
f With the corrective lens in place, calculate the new position of the near point for this eye.

4 a Explain the eye defect *astigmatism*.
b Sketch a chart that might be used to help identify this defect.
c Explain what a person suffering from astigmatism might see when viewing the chart you have drawn.

5 An eye has a near point of 60 cm. For this question you are to assume that all of the refraction in the eye occurs at the front surface of the cornea, and that the distance of the cornea from the retina is 1.8 cm.
a Calculate the power that an eye with normal vision would have when viewing an object at its near point of 25 cm, if the physical dimensions of this normal eye are the same as the eye described at the start of the question.
b Calculate the power of the eye described at the start of the question, when it focuses on an object at its near point of 60 cm.
c State the type of lens needed to correct the sight of the eye and calculate the power of the corrective lens for this eye.

The ear and hearing

In this chapter you will read about:

- the basic structure of the ear and how it detects sound
- how the ear responds to different frequencies and intensities of sound
- what is meant by loudness and the sensitivity of the ear
- some effects of noise

Figure 2.1

Introduction

The detection of sound waves by the ear is known as hearing. Hearing depends on certain characteristics of the sound wave, in particular:

1 the **pitch** of the sound is determined by the frequency of the wave. The normal ear can detect frequencies in a range that varies by a factor of 1000.
2 the perceived **loudness** of the sound is determined by the intensity of the wave and also on its frequency. The typical ear can detect intensities in a range that varies by a factor of 10^{12}.

Not only does the ear have impressive ranges for pitch and loudness, but it also discriminates between sounds according to their **timbre**, or **quality**. This is determined by the number of different frequencies present in a particular sound wave, and their relative amplitudes.

The structure of the ear

The purpose of the ear is to convert small variations in air pressure into electrical signals that are sent to the brain. Figure 2.2 is a simplified diagram of a vertical section through a human ear, showing the main physical features.

In order to study the way in which the ear detects sound waves, it is convenient to divide the ear into three sections: the **external** or **outer ear**, the **middle ear** and the **inner ear**.

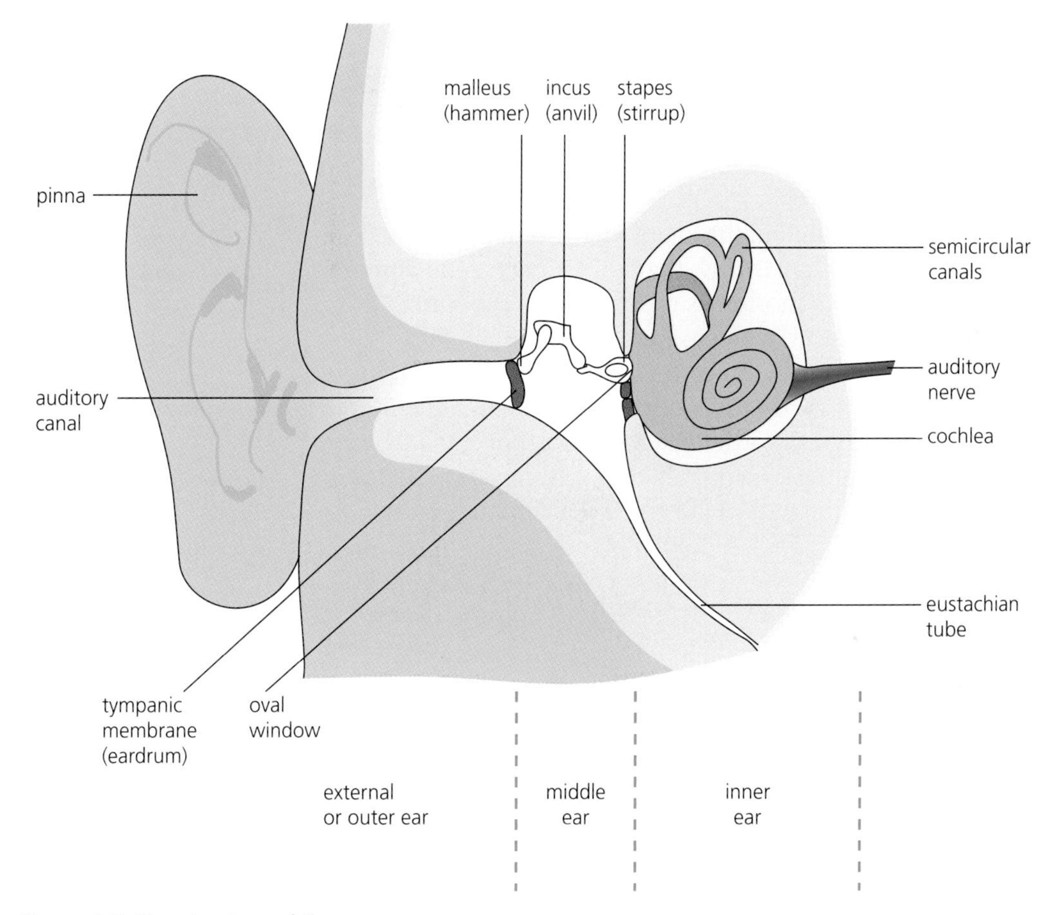

Figure 2.2 The structure of the ear

The external (outer) ear

The visible part of the ear (the **pinna**) plays only a minor role in humans but in some animals its function is to collect sound waves and to direct them towards the **auditory canal**. It also assists with the determination of the direction from which a sound is coming.

The auditory canal is about 2.5 cm long and 7 mm diameter, and leads to the **tympanic membrane** or **eardrum**. The auditory canal acts as a pipe, closed at one end by the tympanic membrane. This membrane has about the thickness of a sheet of paper with an area of about 65 mm^2. Although the membrane is tough, it can be ruptured by sudden loud sounds (explosions) or sudden changes in atmospheric pressure, or even by cotton wool buds. (Remember the old saying – don't put anything in your ear apart from your elbow!) The tympanic membrane (eardrum) is able to respond to air vibrations with amplitudes as small as 10^{-11} m.

The middle ear

The middle ear is an air-filled cavity providing a mechanical connection, or linkage, between the tympanic membrane and a smaller membrane called the **oval window**. This linkage is provided by three small bones, the **ossicles**, that are held in position by ligaments. These bones are referred to as the **malleus** (or **hammer**), the **incus** (or **anvil**) and the **stapes** (or **stirrup**). The ossicles act as a series of pistons and levers, as explained in Box 2.1.

The effect of the ossicles is to increase pressure variations at the oval window but, at the same time, the amplitude of vibration is reduced. The ossicles are said to give rise to 'impedance matching', that is, the maximum amount of sound power incident on the tympanic membrane is transferred to the inner ear with as little as possible being reflected at the oval window.

The **eustachian tube** links the middle ear to the back of the mouth. This enables the ambient pressure on each side of the tympanic membrane to remain equal. Any differences in pressure, other than those associated with normal sound waves, cause the tympanic membrane to stretch, resulting in discomfort or pain. Swallowing, yawning or chewing will usually open the eustachian tube, allowing the pressures to equalise. You may have experienced this effect when in a plane or a lift. In children, the eustachian tube may be narrow or blocked, causing earache. 'Grommets' may be inserted through the tympanic membrane to relieve any build-up of pressure in the middle ear.

The inner ear

The inner ear is a small cavity within the bone of the skull that is filled with a liquid known as perilymph. Within this inner ear are two organs. The **semicircular canals** are liquid-filled tubes that are not part of the hearing mechanism. Instead, they detect movement of the body and assist with the regulation of balance. The **cochlea** is a spiral tube, also filled with liquid. It has a total volume of about 100 mm^3 and, if it were to be uncoiled, it would be about 3 cm long.

The functioning of the cochlea is a complex process that is not, as yet, fully understood. Running along its length are two membranes that divide the cochlea into three cavities. One of these membranes, the **basilar membrane**, is composed of many hairs of different lengths and stiffnesses. The lengths vary from about 0.04 mm to about 0.5 mm. The different lengths and stiffnesses mean that the hairs will all **resonate** (be set into vibration) at different frequencies. When a pressure wave enters the cochlea via the oval window, certain hairs vibrate, dependent upon the frequencies of the incident vibrations. These vibrating hairs cause small electrical signals that are communicated to the brain via nerve fibres.

Box 2.1 The action of the ossicles

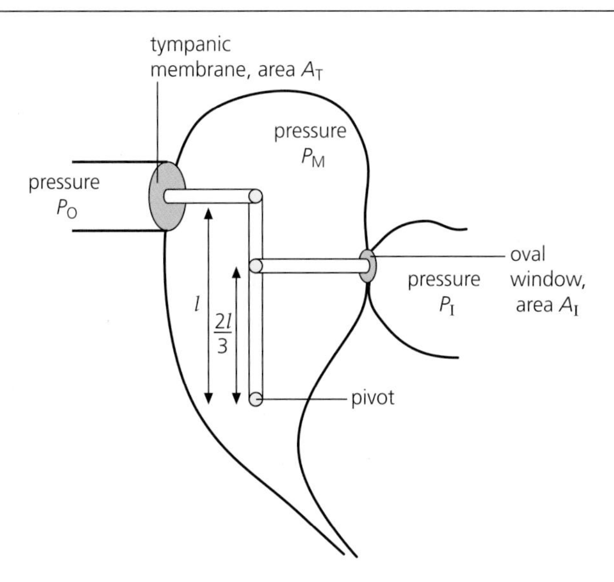

Figure 2.3 The lever system of the middle ear

Consider a pressure difference $(P_O - P_M)$ between the outer ear and the middle ear, that is, across the tympanic membrane (see Figure 2.3). If the area of the tympanic membrane is A_T, then the resultant force F_T on the malleus is given by:

$$F_T = A_T(P_O - P_M)$$

The force on the malleus is transmitted through to the oval window by means of the incus and stapes. The ossicles provide a lever system so that, using the principle of moments, the force F_I acting on the oval window is given by:

$$F_I \times \tfrac{2}{3}l = F_T \times l$$
$$F_I = \tfrac{3}{2}F_T$$

The pressure difference $(P_I - P_M)$ between the inner ear and the middle ear is given by:

$$(P_I - P_M) = \frac{F_I}{A_I}$$

where A_I is the area of the oval window. This gives the pressure difference ratio:

$$\frac{(P_I - P_M)}{(P_O - P_M)} = \frac{A_T}{A_I} \times \frac{F_I}{F_T}$$

Now $A_T/A_I \approx 20$ (the area of the oval window is about $3\,\text{mm}^2$) and $F_I/F_T = \tfrac{3}{2}$. Therefore:

$$\frac{(P_I - P_M)}{(P_O - P_M)} \approx \tfrac{3}{2} \times 20 = 30$$

The response of the ear

The ear is a delicate mechanism that can respond to a wide range of frequencies and intensities. Many of us take for granted the ability to hear. It is only when we lose that ability that we appreciate what we have lost. Even for people with normal hearing, there are limits to the frequencies and intensities of sound that can be heard.

Frequency response

As outlined on page 28, the frequency response of the ear is based on **resonance**. Short stiff hairs in the cochlea resonate at high frequencies, whereas the longer fibres respond at low frequencies. Below about 20 Hz, there is no stimulation of the hairs and this gives rise to a lower limit of frequency that can be detected. The upper limit is about 20 kHz although this limit becomes lower with age. Thus the range of audible frequencies is about 20 Hz to 20 kHz.

The upper limit of the frequency range is different for different animals. Dogs can detect frequencies above 20 kHz and this forms the basis of the dog whistle. In general, frequencies above 20 kHz are called **ultrasonic** frequencies.

Resonance also occurs in regions of the ear other than the cochlea. The auditory canal acts as a tube of length about 2.5 cm, closed at one end. Thus a standing wave may be set up in this canal with a displacement node near the tympanic membrane and an antinode near the pinna. The wavelength of this standing wave will be about $4 \times 2.5 = 10$ cm (Figure 2.4). If a sound wave of this wavelength enters the ear, resonance will occur in the canal, giving rise to greater pressure differences across the tympanic membrane.

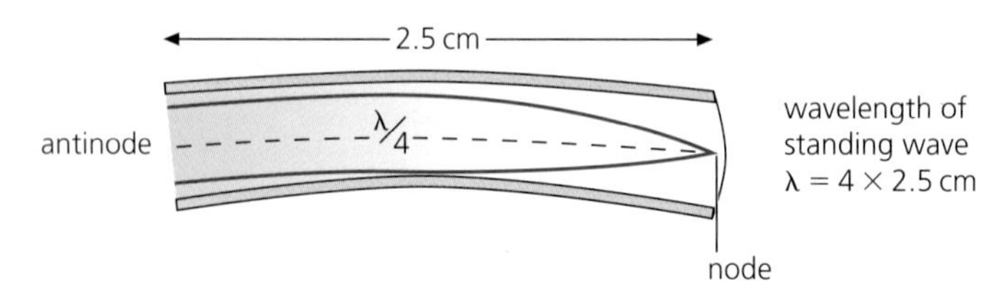

Figure 2.4 Resonance in the auditory canal

Since the speed of sound in air is about $330\,\text{m}\,\text{s}^{-1}$, the resonant frequency f is given by:

$$f = \frac{\text{speed}}{\text{wavelength}}$$

$$= \frac{330\,\text{m}\,\text{s}^{-1}}{0.1\,\text{m}} = 3300\,\text{Hz}$$

The middle ear also acts as a resonant cavity, giving resonances between about 700 Hz and 1500 Hz.

These resonances mean that the required intensity of a sound wave incident on the ear, such that a particular frequency can be detected, is not constant. Because of resonance in the auditory canal, lower intensities can be heard for frequencies around 3 kHz. Much research has been carried out into measuring the lowest intensity that can be detected by people with normal hearing. The results are illustrated in Figure 2.5.

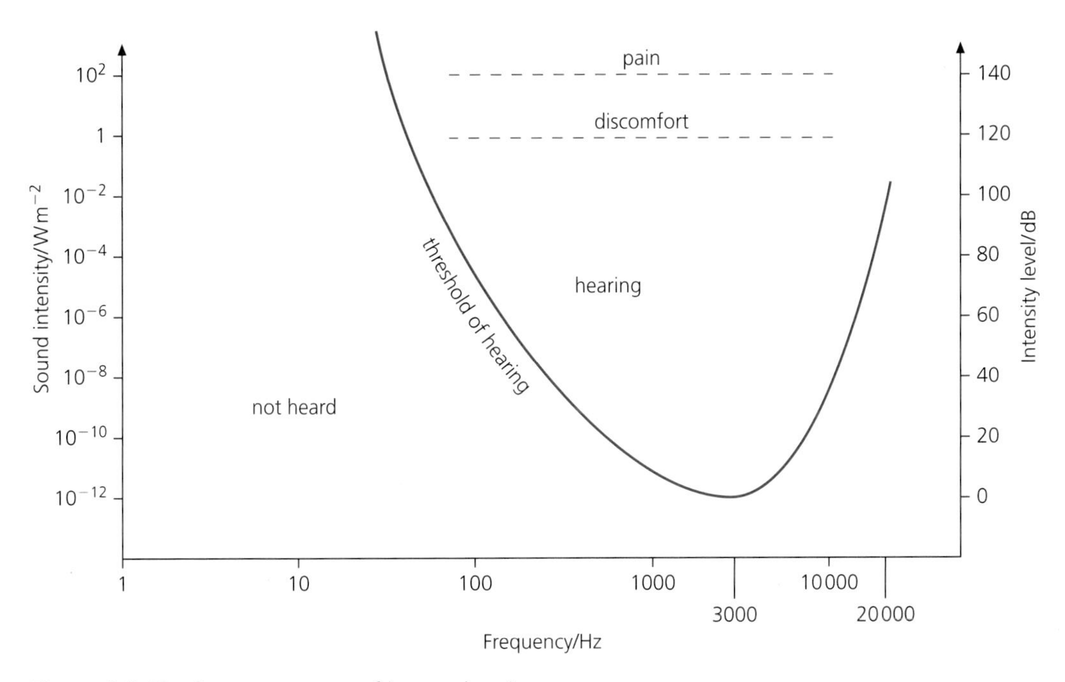

Figure 2.5 The frequency range of human hearing

Note that the *x*-axis and the *y*-axis scales have been plotted logarithmically. The use of **intensity level**, measured in **decibels** (dB), will be discussed later in the chapter (page 34).

The line on the graph represents the **threshold of hearing**, that is, the minimum or 'threshold' intensity incident on the ear such that sound of a particular frequency can be detected by a person with normal hearing.

It can be seen that the range of frequencies that can be detected depends on the intensity of sound. In general, higher intensities give rise to a larger range of audible frequencies. However (be warned!) very high intensities can cause discomfort, pain, temporary or even permanent deafness.

Within the range of audible frequencies, the ear can detect changes of frequency. For low frequencies in the range 60 Hz to 1000 Hz, differences as small as 2 or 3 Hz can be distinguished when the frequencies are sounded separately. This ability to distinguish between frequencies becomes less as the frequency increases until, at about 10 kHz, discrimination is very poor. This fall-off in discrimination with increase in frequency is mirrored in the development of musical scales. Each octave corresponds to a doubling of frequency, not an equal difference, e.g. middle C is 256 Hz, upper C is 512 Hz and top C is 1024 Hz.

Intensity response

As discussed in the previous section, the minimum sound intensity at any particular frequency that can be detected by the ear is known as the **threshold intensity** for that particular frequency. This threshold intensity is frequency dependent. Its lowest value is about $10^{-12}\,W\,m^{-2}$ at 2–3 kHz (see Figure 2.5, page 31). Since the area of cross-section of the auditory canal is about $0.4\,cm^2$, this lowest value corresponds to a sound power of only $(0.4 \times 10^{-4} \times 10^{-12}) = 4 \times 10^{-17}\,W$. Interestingly, if the threshold were to be much lower, the thermal motion of air molecules would be detected. Furthermore, if the sensitivity of the ear did not drop off at frequencies below about 3 kHz, we would hear our own blood flowing through the arteries.

Whenever the intensity of sound increases, the brain perceives this as an increase in **loudness**. An increase in loudness is brought about as a result of:

1 greater stimulation of nerve endings due to an increased amplitude of vibration of hairs in the cochlea
2 more nerve endings being stimulated because the amplitude of vibration of all hairs is increased
3 nerve cells having higher thresholds for stimulation being activated.

Loudness is, therefore, a complex reaction to sound of a particular intensity. Also, what may seem loud to one person may not for another. Loudness is a subjective response, whereas **intensity** is a defined physical quantity. It is the energy of the sound wave arriving at the ear per second per m^2 of cross-sectional area. The intensity of sound at the ear produced by some sources is shown in Table 2.1.

Table 2.1 Sound sources and intensities at the ear

Sound source	**Sound intensity at ear** /W m^{-2}
Threshold of hearing	10^{-12}
Rustling leaves	10^{-10}
Background music	10^{-8}
Speech	10^{-6}–10^{-4}
Noisy factory	10^{-3}
Aircraft overhead	10^{-2}
Thunder overhead	10^{-1}
Noise causing discomfort	1

Notice the very wide range of intensities that can be detected – a range covering 12 orders of magnitude. With such a wide range, it is to be expected that the response of the ear is not linear. In fact, at low intensities, only small changes in intensity are required for a change in noise level to be noticed. At high intensities, large changes are required before a change is detected. In other words, equal changes in intensity across the audible range are not perceived as equal changes in loudness. What is important is the *ratio* of the change in intensity to the original intensity. Equal ratios are detected as equal changes in loudness. That is,

$$\text{change in loudness depends on } \frac{\text{intensity change}}{\text{initial intensity}}$$

Look at the intensities shown in Table 2.2.

Table 2.2

Intensity I /W m^{-2}	**Loudness perceived**	$\frac{I}{I_0} = \frac{\text{intensity}}{1.0 \times 10^{-12}}$	$\lg \frac{I}{I_0}$
1.0×10^{-12} (I_0)	0	1	0
2.0×10^{-12}	L	2	0.30
4.0×10^{-12}	$2L$	4	0.60
8.0×10^{-12}	$3L$	8	0.90

The first intensity has been chosen as 1.0×10^{-12} W m^{-2} because this is the lowest intensity that can be detected, i.e. it is the threshold of hearing, termed I_0. The subsequent intensities each show a doubling of intensity and would be detected by the ear as equal changes in loudness. If one step in loudness is represented by L, then the loudness would be zero at the threshold of hearing, L at 2.0×10^{-12} W m^{-2}, $2L$ at 4.0×10^{-12} W m^{-2}, etc.

Dividing each value of intensity I by the threshold intensity I_0 gives the numbers 1, 2, 4, etc. What is then of significance is if the logarithm of each ratio is calculated. Looking at Table 2.2 again, you can see that the logarithm (lg) increases in equal steps, following the equal changes in loudness. We can conclude that the response of the ear to sound intensity is **logarithmic**.

This logarithmic response means that it is possible to obtain a measure of the response of the ear to a sound of intensity I by calculating what is called the **intensity level** (IL):

$$\text{intensity level (IL)} = \lg\frac{I}{I_0}\ \text{bels} \quad (\text{where } \lg = \log_{10})$$

I_0 is the sound intensity at the threshold of hearing and is taken as $1.0 \times 10^{-12}\,\mathrm{W\,m^{-2}}$. Intensity level (IL) is measured in **bels**, given the symbol B. This is a large unit and, in general, intensity levels are expressed in **decibels** (dB), where $10\,\mathrm{dB} = 1\,\mathrm{B}$. Therefore:

$$\text{intensity level in dB} = 10\lg\frac{I}{I_0}$$

The intensity levels corresponding to some intensities are given in Table 2.3.

Table 2.3 Intensity levels of some sounds

Sound	**Intensity** /$\mathrm{W\,m^{-2}}$	**Intensity level** /dB
Threshold of hearing	10^{-12}	0
Whispering	10^{-9}	30
Quiet classroom	10^{-7}	50
Electric drill at 3 m	10^{-5}	70
Heavy traffic at 4 m	10^{-3}	90
Club music	10^{-1}	110
Threshold of pain	10^{2}	140

WORKED EXAMPLE 2.1

A noisy food mixer produces sound of intensity $8.3 \times 10^{-5}\,\mathrm{W\,m^{-2}}$ at the ear.

a Calculate the *intensity level* (IL) of this noise at the ear.

b At this level of noise, the minimum change in intensity level that can be detected by the ear is 3 dB. Determine the change in intensity corresponding to this change in intensity level.

a

$$\mathrm{IL} = 10\,\mathrm{lg}\frac{I}{I_0}$$

$$= 10\,\mathrm{lg}\frac{8.3 \times 10^{-5}}{1.0 \times 10^{-12}}$$

$$= 79.2\,\mathrm{dB}$$

b Either consider an intensity level of $(79.2 + 3)$ dB:

$$(79.2 + 3) = 10\,\mathrm{lg}\frac{I}{1.0 \times 10^{-12}} \quad \text{where } I \text{ is the new, increased intensity}$$

giving

$$I = 1.66 \times 10^{-4}\,\mathrm{W\,m^{-2}}$$

and so

$$\text{change in intensity} = (1.66 \times 10^{-4}) - (8.3 \times 10^{-5})$$
$$= 8.3 \times 10^{-5}\,\mathrm{W\,m^{-2}}$$

Or consider the *change* in intensity level of +3 dB:

$$3 = 10\,\mathrm{lg}\frac{I}{I_0} - 10\,\mathrm{lg}\frac{8.3 \times 10^{-5}}{I_0}$$

$$= 10\,\mathrm{lg}\frac{I/I_0}{8.3 \times 10^{-5}/I_0}$$

$$= 10\,\mathrm{lg}\frac{I}{8.3 \times 10^{-5}}$$

giving

$$I = 1.66 \times 10^{-4}\,\mathrm{W\,m^{-2}}$$

and

$$\text{change in intensity} = (1.66 \times 10^{-4}) - (8.3 \times 10^{-5})$$
$$= 8.3 \times 10^{-5}\,\mathrm{W\,m^{-2}}$$

Note that a 3 dB change in intensity level corresponds to a doubling of the actual intensity.

WORKED EXAMPLE 2.2

The sound power of the explosion of a firework is 2.5 W. This sound power is distributed uniformly in all directions over a spherical area that expands outwards from the source at the speed of sound. Calculate, for a person standing at a distance of 50 m from the firework:

a the intensity of the sound at the ear,
b the intensity level.

a $\text{Intensity} = \dfrac{\text{power}}{\text{area}}$

The area we need is the surface area of the sphere whose radius is the distance of the person from the firework, so

$$\text{intensity } I = \frac{\text{power}}{4\pi r^2}$$

$$= \frac{2.5\,\text{W}}{4\pi(50)^2\,\text{m}^2}$$

$$= 7.96 \times 10^{-5}\,\text{W m}^{-2}$$

b Intensity level is given by:

$$\text{IL} = 10\lg\frac{I}{I_0}$$

$$= 10\lg\left(\frac{7.96 \times 10^{-5}}{10^{-12}}\right)$$

$$= 79\,\text{dB}$$

Loudness and sensitivity

Loudness is the ear's *subjective* response to sound intensity but it varies in the same way as intensity level. That is, the ear has a logarithmic response. Loudness is frequency dependent at any particular intensity, as illustrated in Figure 2.6.

Each curve on the graph represents equal loudness. Although loudness is a subjective response, a quantitative measure can be made. Loudness is measured in **phon**.

In order to measure the loudness of a sound at a particular frequency, the sound under test is compared with a sound of frequency 1 kHz. The loudness of the sound at 1 kHz is varied until it is perceived to have the same loudness as the sound to be measured. The loudness of the sound (measured in phon) is then numerically equal to the intensity level of the 1 kHz sound wave. Thus, any line of equal loudness has a numerical value, for the number of phon, equal to its intensity level at 1 kHz.

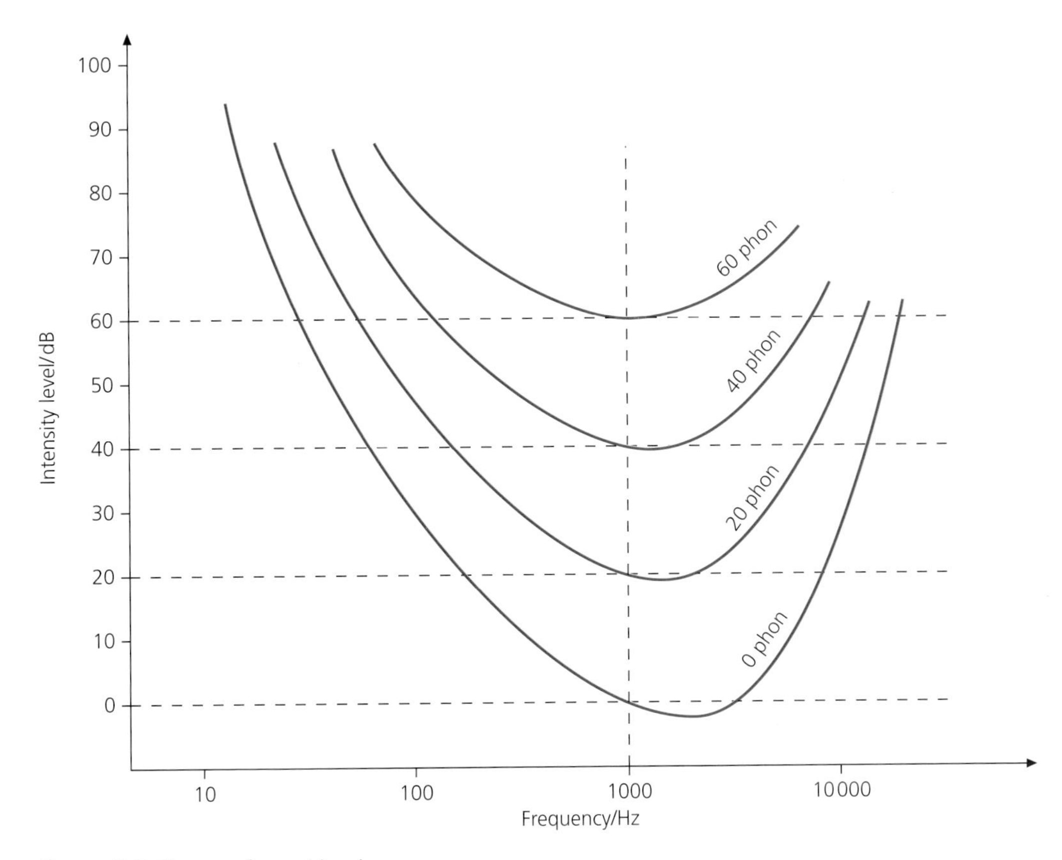

Figure 2.6 Curves of equal loudness

When the ear is able to detect changes in loudness or intensity level, the ear is said to be sensitive to change. The **sensitivity** S of the ear is its ability to detect the smallest fractional change ($\Delta I/I$) in intensity. It is defined by the equation:

$$S = \lg \frac{I}{\Delta I}$$

Sensitivity is strongly dependent on frequency and intensity. For a typical intensity, sensitivity varies as illustrated in Figure 2.7 overleaf.

Note that the maximum sensitivity S_{max} occurs at about 2 kHz. The relative sensitivity S/S_{max} has been plotted on the y-axis, where S is the actual sensitivity at the frequency and intensity being considered.

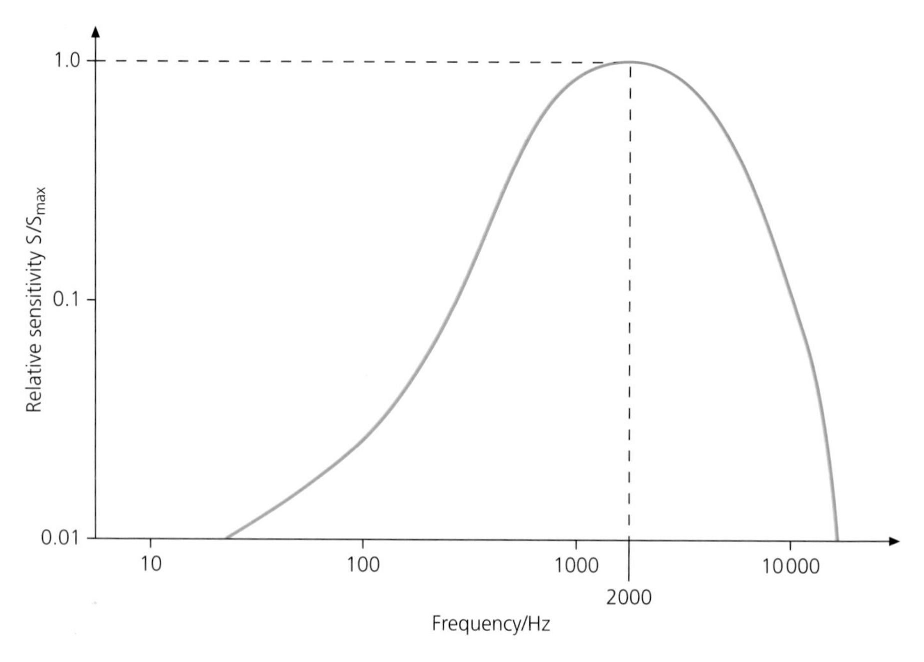

Figure 2.7 Variation in the ear's sensitivity

Typically, the minimum change in intensity level that can be detected is 0.5 dB. The corresponding ratio of sound intensities I/I_1 is given by:

$$0.5 = 10\,\lg\frac{I}{I_1}$$

$$\frac{I}{I_1} = 1.12$$

That is, the ear is typically sensitive to a 12% change in intensity.

The effect of noise

Noise is, in general, sound with many different frequency components, all with different and varying amplitudes. Noise causes different effects in different individuals. Noise that may, to some people, be intensely annoying will seem to be acceptable to others. The total absence of any sound, or noise, can also be very disturbing.

It is now generally accepted that harmful effects of noise become more severe as the noise level increases. What can be termed as unacceptable noise may be as low as 85 dB.

Hearing naturally deteriorates with age. This is called **presbycusis**. The green curves in Figure 2.8 illustrate the variation with frequency of the loss of hearing, measured in dB, of two people aged 40 and 65.

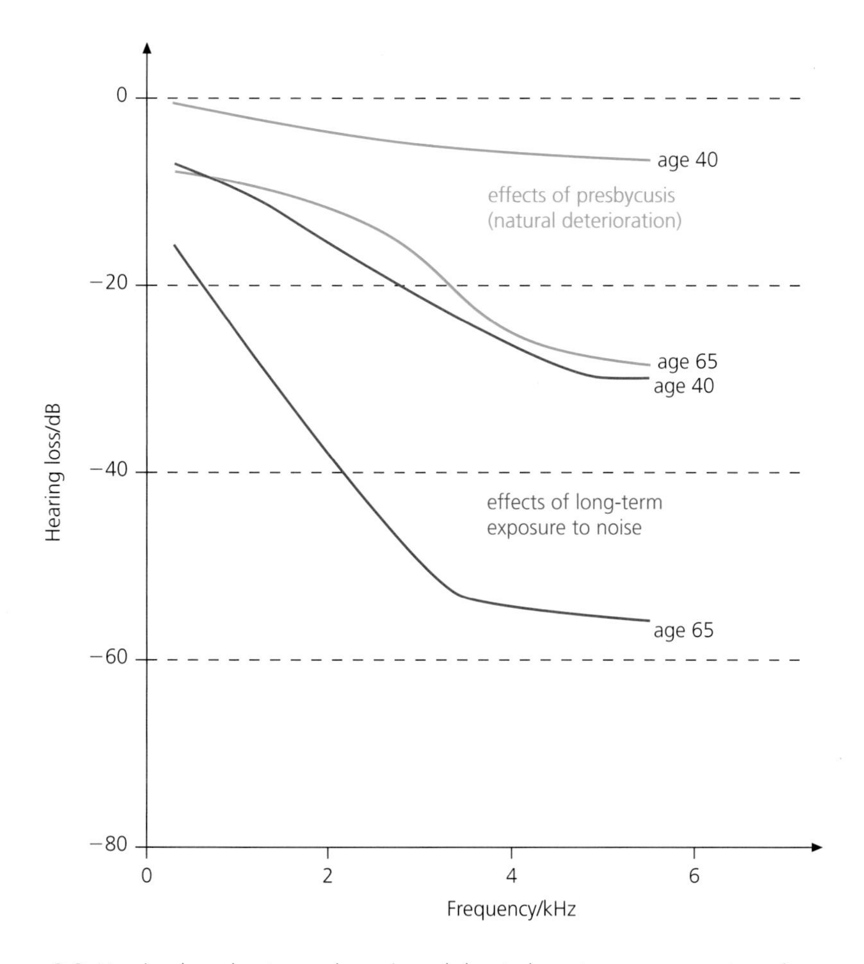

Figure 2.8 Hearing loss due to presbycusis and due to long-term exposure to noise

Neither of these people represented by the green curves has been exposed to noisy surroundings. Note how the higher frequencies are most affected. A negative number on the dB scale represents a loss of hearing. That is, −20 dB is a hearing loss of 20 dB, meaning that the intensity level for a sound to be heard would need to be 20 dB greater than for a person with 'normal' hearing.

The harmful effects of long-term exposure to high noise levels are illustrated by the red curves. These show typical hearing loss of two people aged 40 and 65 who have been exposed to noise at a level of 95 dB during their working days since leaving school. It can be seen quite clearly why legislation has been introduced to limit noise in the workplace. Where noise cannot be eliminated, then earplugs and/or earmuffs must be worn. Health hazard notices should perhaps be shown at discos and clubs!

Besides the long-term effects of hearing loss, high noise levels may cause temporary or permanent **tinnitus**. This is a ringing noise in the ears, which makes it difficult to understand what people are saying, particularly if there is any background noise.

Other effects of noise include:

1 feelings of annoyance
2 inability to think clearly
3 dizziness or sickness (>125 dB)
4 pain in the ears (>130 dB)
5 permanent deafness (~190 dB, short-term exposure)

Questions

1 a Draw a labelled diagram to show the main physical features of the human ear.
b By reference to your diagram, describe the processes that take place during the detection of a sound wave by the ear.

2 a Outline briefly how the ear is able to distinguish sounds of different frequencies.
b i) State the range of frequencies detectable by the ear for a person with normal hearing.
ii) Suggest why there is an upper and a lower limit to the frequencies that can be detected.
iii) Describe the effect on the audible frequency range of the ageing process.

3 a i) Sketch a graph to show how the threshold of hearing varies with frequency. On your graph, mark any significant values of frequency and intensity level.
ii) With reference to Figure 2.8, draw a second line on your graph to show how the threshold of hearing might vary for a person who is 90 years of age.
b The threshold of hearing at a frequency of 3 kHz for a person with defective hearing is $3.2 \times 10^{-10}\,\mathrm{W\,m^{-2}}$. Calculate the loss, in dB, of the person's hearing.

4 a Explain why a logarithmic scale is used in order to quantify the ear's response to sound intensity.
b A loudspeaker may be considered to be a point source of sound emitting sound power uniformly in all directions. The sound power produced is 0.90 W. Calculate, for a point 1.5 m from the loudspeaker,
i) the intensity of sound,
ii) the intensity level.
The surface area A of a sphere of radius r is given by $A = 4\pi r^2$.
c Comment on your answer to **b** ii) with reference to any possible health hazard.

5 a An aircraft passes overhead. The maximum intensity of sound experienced by an observer is $7.4 \times 10^{-3}\,W\,m^{-2}$. Calculate the equivalent intensity level.

b The intensity level at a disco is 95 dB. Calculate the intensity of sound in the room.

c The mean intensity level by the side of a road is 68 dB. When a heavy lorry passes, the intensity level rises to 92 dB. Determine the fractional change in intensity corresponding to this change in intensity level.

6 a i) Define what is meant by the *sensitivity* of the ear.

ii) Describe how the sensitivity of the ear varies with frequency.

b The minimum change in intensity level that a person can detect for a background noise level of 85 dB is 2.5 dB. Determine the fractional change in intensity corresponding to this sensitivity.

The cardiovascular system

In this chapter you will read about:

- aspects of nerve cells and biopotentials
- the principles of the artificial pacemaker and the defibrillator
- blood circulation and blood pressure

Biopotentials

As long ago as the 18th century, it was realised that there is a connection between the action of muscles and electric potentials. In 1786, Galvani showed that a frog's leg could be made to twitch by applying a suitable potential difference across it. However, the ability to measure potentials associated with the action of muscles was not achieved until the beginning of the 20th century. This ability, together with the development of microelectronics, has led to important discoveries about the functioning of the body. The measurement of potential differences created within the body, or **biopotentials**, is now used routinely in medical diagnosis.

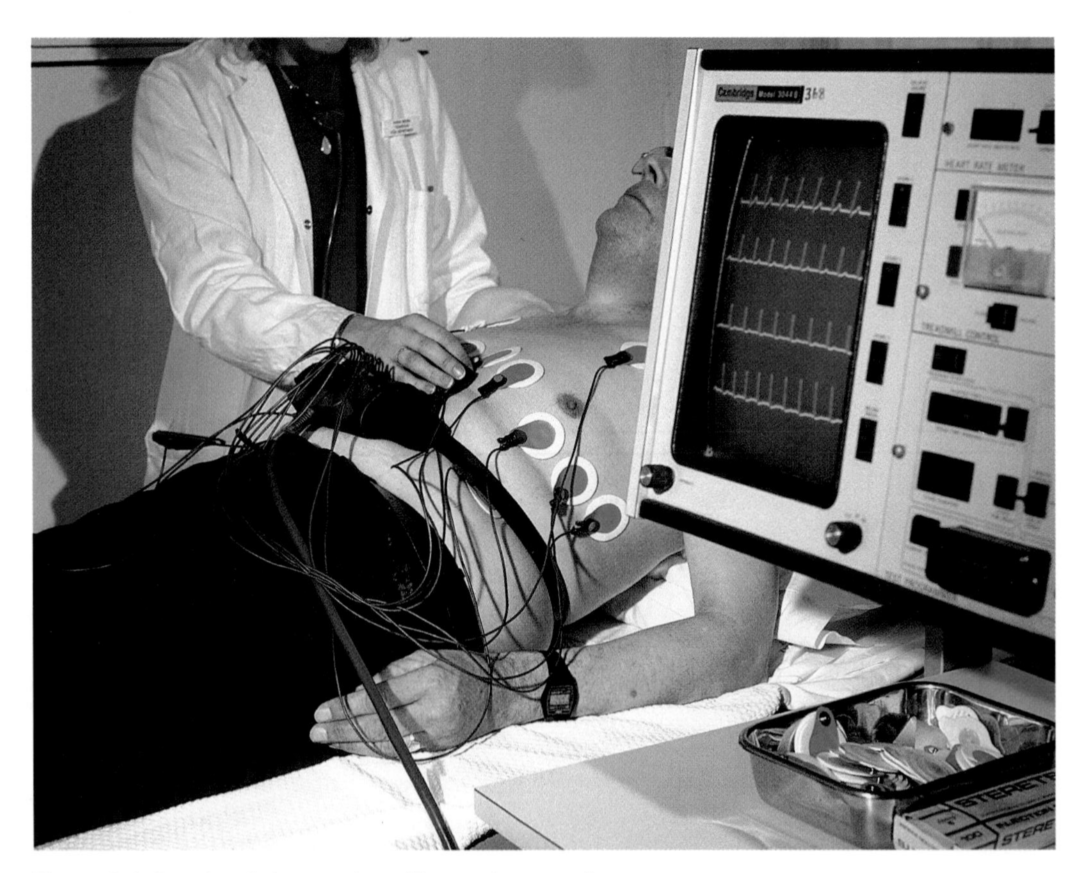

Figure 3.1 A patient being monitored by an electrocardiogram

Nerve cells

One of the communication systems within the body is the **nervous system** which is made up of nerve cells. A typical nerve cell (**neuron**) for the control of movement is illustrated in Figure 3.2.

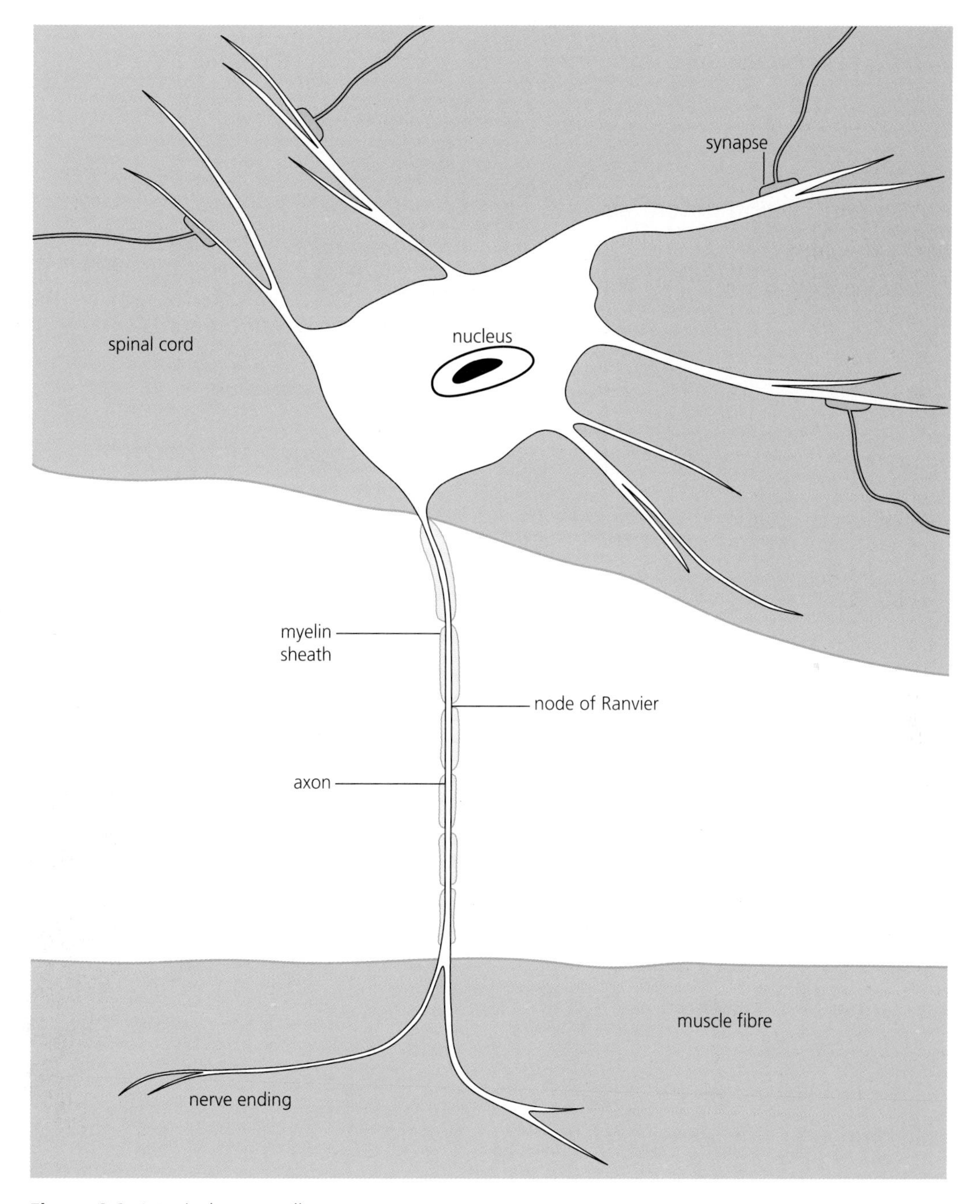

Figure 3.2 A typical nerve cell

The body of the cell containing the nucleus is found in the spinal column. Messages are sent to the muscle along thin fibres called **axons**. An axon is, typically, a few micrometres in diameter and may be more than a metre in length. In order to insulate the axon from surrounding tissues, it is covered with a **myelin** sheath. At intervals of about 1 mm along the axon, there are gaps in the sheath. These gaps, called the **nodes of Ranvier**, enable charged ions to flow into or out of the axon. These ions can either amplify or attenuate the electrical signal passing down the axon.

Nerve cells are connected together via **synapses** between the branching ends of the nerve cells.

Resting and action potentials

The nerve fibre (axon) consists of a central core of **axoplasm** surrounded by a high-resistance semi-permeable membrane, as illustrated in Figure 3.3.

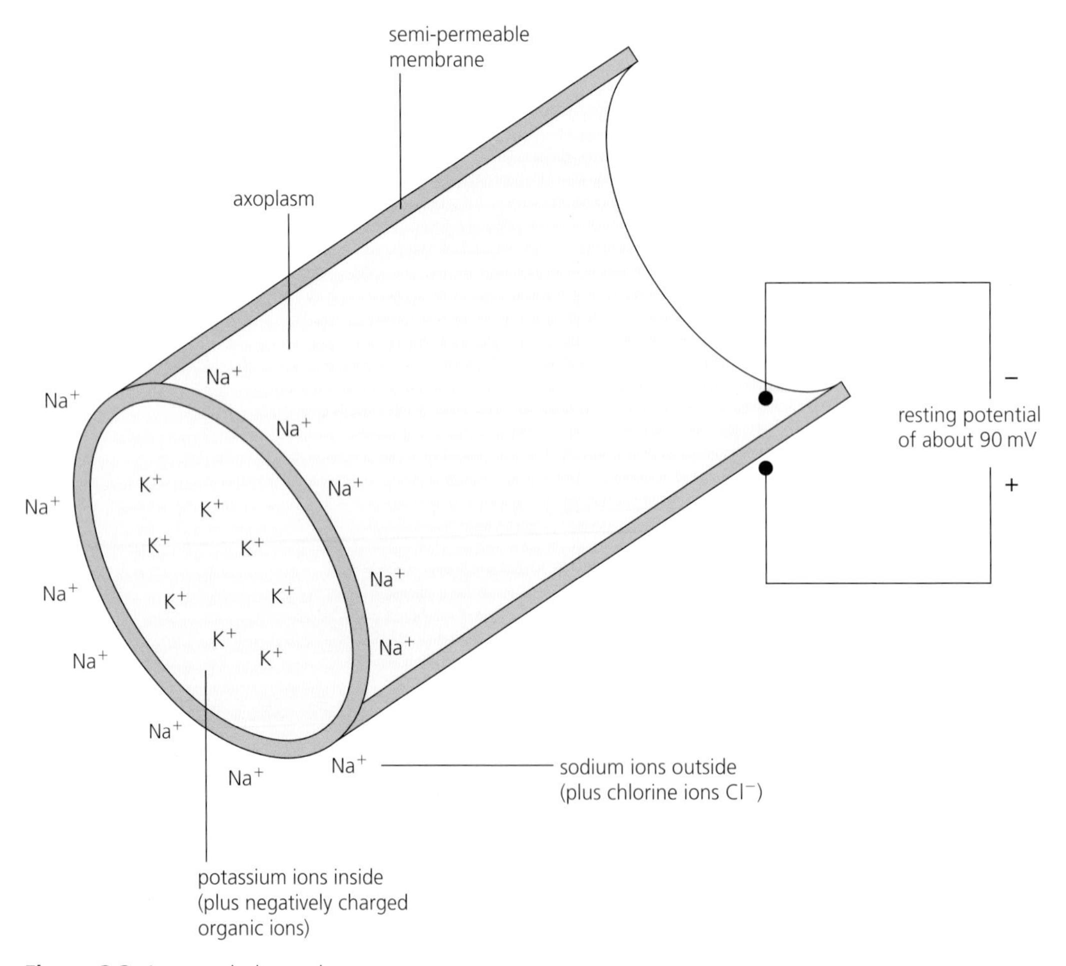

Figure 3.3 An axon in its resting state

The small ions of potassium (K^+), sodium (Na^+) and chlorine (Cl^-) can pass through this membrane. When the nerve is in its resting state, there is a high concentration of K^+ and negatively charged organic ions within the axon. Outside, in the **interstitial fluid**, there is an excess of Na^+ and Cl^- ions. The imbalance is maintained by what is called (but not fully understood) the **sodium–potassium pump**. The result of this imbalance is an excess of positive charge outside the axon and a potential difference of about 90 mV across the membrane. This potential difference is known as the **resting potential**.

When the nerve ending is stimulated, the axon membrane suddenly becomes permeable to Na^+ ions and these move into the axoplasm. This causes the potential difference across the membrane to become smaller (**depolarisation**) and then to reverse in direction (**reverse polarisation**). The process takes only a few milliseconds and continues until the reverse potential difference is about 20 mV. This potential difference is called the **action potential**.

The variation with time of the potential difference across the axon membrane is a characteristic of any particular type of nerve. A typical example is shown in Figure 3.4.

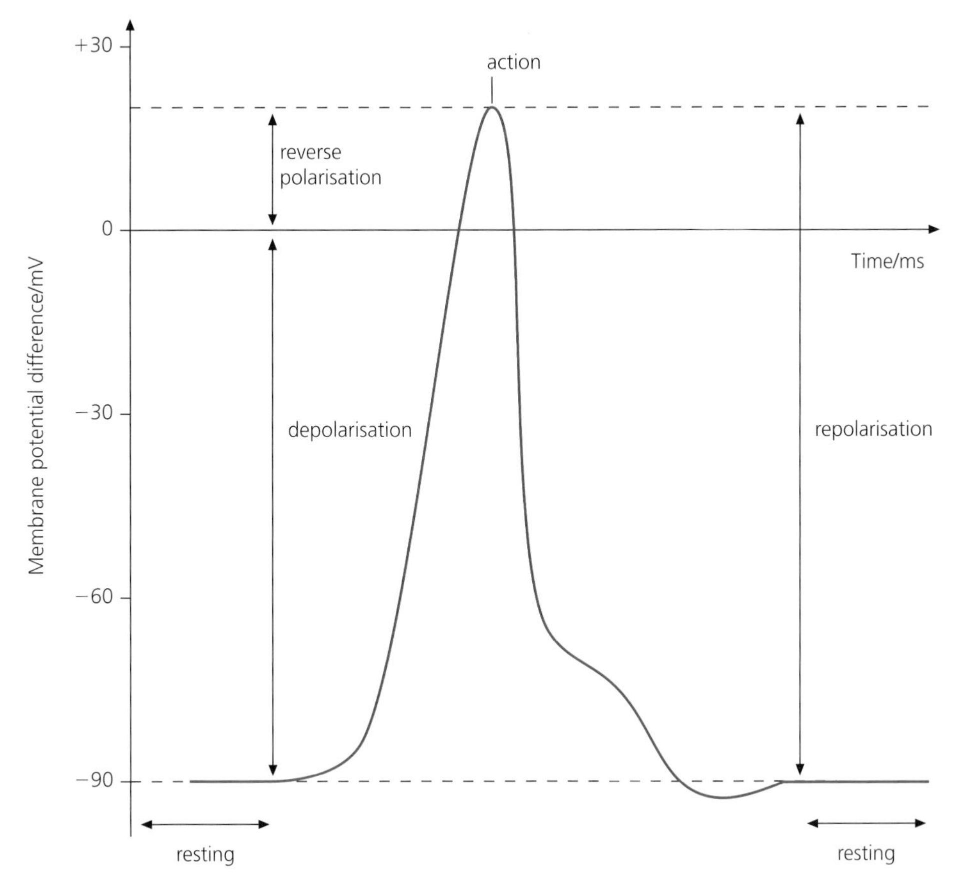

Figure 3.4 Resting and action potentials

As soon as the action potential is reached, the membrane becomes less permeable to Na^+ ions and more permeable to K^+ ions. Over a period of about 50 ms, the sodium–potassium pump restores the axon to its resting potential (**repolarisation**). The axon is then ready to respond to another stimulus.

The depolarised (active) region in the fibre acts as a trigger and stimulates the neighbouring region of the fibre to follow through the same action potential. The action potential moves along the fibre at a speed that depends on the type of cell and the temperature. This speed may be as high as $150\,m\,s^{-1}$ in some nerve cells.

The heart

The circulation of the blood around the body by means of the heart – the **cardiovascular system** – is a vital system for life. If the heart fails, death occurs within a very short time. The heart consists of groups of muscles in which biopotentials are produced. The monitoring of these biopotentials is one means by which the state of health of the heart may be monitored.

The action of the heart

The heart acts as a double pump consisting of four chambers with valves. A schematic diagram illustrating the action of the heart is shown in Figure 3.5.

Blood is received into the right auricle (or atrium) from the body and is then passed through a valve into the right ventricle. From here, the blood is pumped to the lungs. The oxygenated blood returning from the lungs flows into the left auricle (or atrium). The blood then passes through a valve into the left ventricle, from where it is pumped around the body, passing out of the heart through the **aorta**.

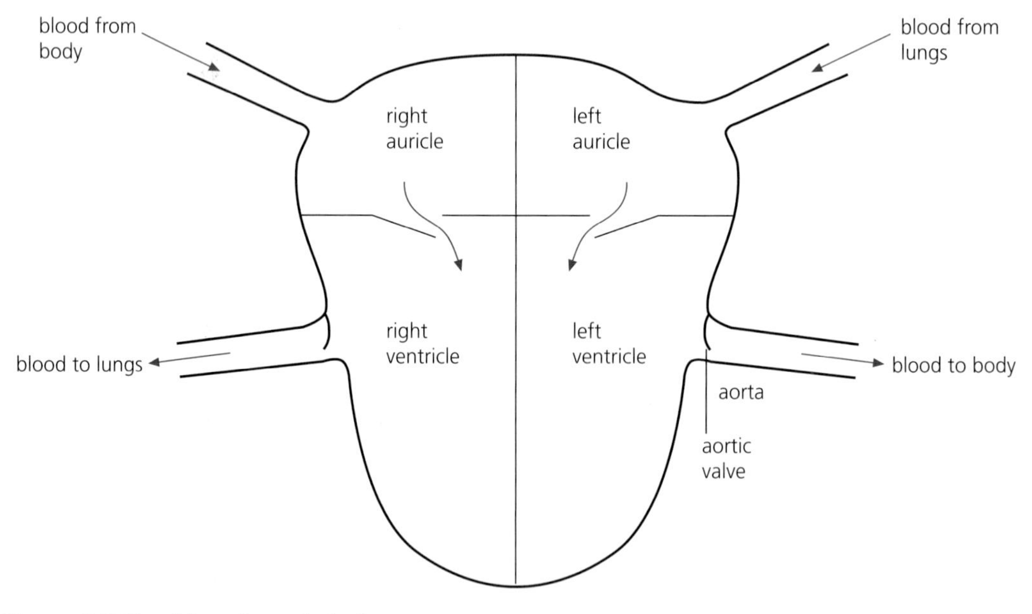

Figure 3.5 Blood flow through the heart

Each beat of the heart is triggered by an electrical pulse from the **sino-atrial (SA) node** or **pacemaker**, located in the right auricle (atrium). The pulse spreads across the auricles (atria), depolarising them and causing them to contract, so forcing blood into the ventricles. Shortly after, the electrical signal passes to the **atrio-ventricular (AV) node**. This triggers the depolarisation of the two ventricles, causing them to contract and so forcing blood out of the heart.

The electrocardiogram (ECG)

During each heartbeat, the spread of the electrical pulse across the heart causes potential differences between polarised and depolarised cells. These potential differences are conducted to the skin through the body fluids and make up an electrical signal that can be detected by suitable electrodes placed on the skin. The signals are usually only a few millivolts in amplitude but they may be amplified and displayed as an **electrocardiogram (ECG)**. A typical ECG waveform is illustrated in Figure 3.6.

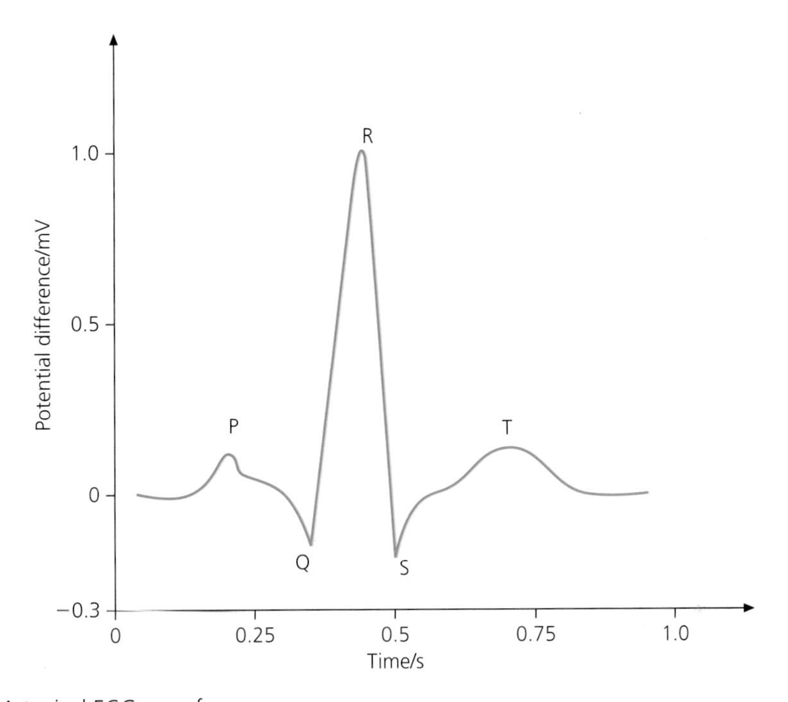

Figure 3.6 A typical ECG waveform

On the waveform, three important features have been labelled. These are the P-wave, the QRS-wave and the T-wave.

1 The P-wave occurs during depolarisation of the auricles, causing them to contract.
2 The QRS-wave corresponds to depolarisation and contraction of the ventricles.
3 The T-wave results from repolarisation and relaxation of the ventricles.

The ECG is usually recorded, after amplification, on a chart recorder or displayed on a cathode ray oscilloscope. The waveform measured at the body's surface depends not only on the individual person but also on the siting of the electrodes. Figure 3.7 shows some different waveforms produced by the same heartbeat by electrodes placed on different parts of the body.

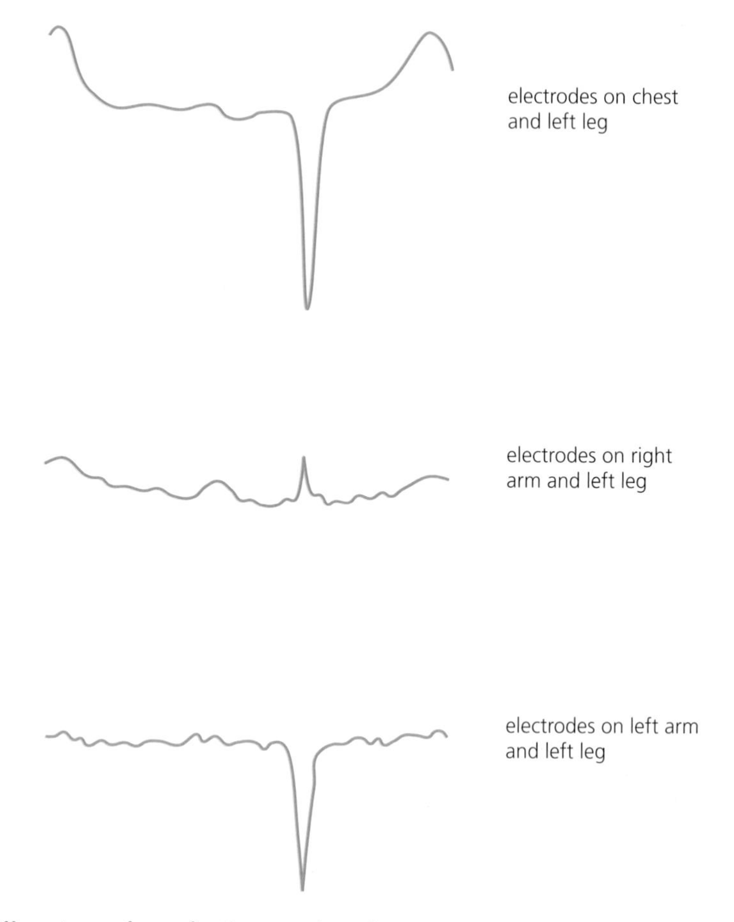

Figure 3.7 Different waveforms for the same heartbeat

The shape of the signal at any one location will depend on the state of health of the heart muscles. Consequently, the study of an ECG gives diagnostic information about the heart. Some common cardiac disorders that can be diagnosed with an ECG are:

- high pulse rate (**tachycardia**)
- low pulse rate (**bradycardia**)
- ventricular fibrillation – irregular contractions of heart muscles
- damaged heart muscle – wave heights are reduced
- heart blockage – part of the trace is missing.

The artificial pacemaker

As described on page 47, the pumping action of the heart is controlled by the sino-atrial and atrio-ventricular nodes. For a resting adult, the heart has a pulse rate of about 70 per minute. This pulse rate is controlled so that, at times of exertion, it increases to meet the oxygen demands of the body. If the atrio-ventricular node becomes damaged, the heart does not stop but instead it goes into automatic control at a pulse rate of about 30 per minute. This rate is sufficient to maintain life, but only if the person is very inactive.

In order to improve this situation, artificial pacemakers have been developed. The artificial pacemaker provides electrical pulses directly to the heart and replaces the action of the atrio-ventricular node.

The pacemaker is inserted into the body (possibly in the abdominal cavity or under the shoulder, Figure 3.8) and the wire is fed through a vein to the appropriate position in the heart. The batteries powering the pacemaker last for several years. In early types of artificial pacemaker, the electrical pulse rate to the heart could not be varied and was fixed at about 70 per minute. Modern developments mean that the pacemaker can change its pulse rate, dependent upon the oxygen demands of the body.

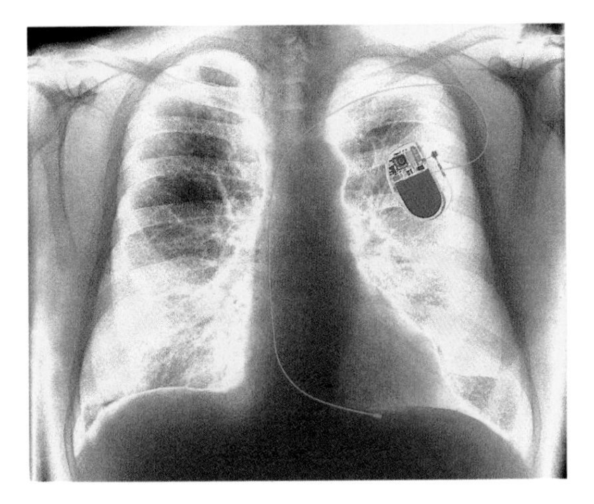

Figure 3.8 An artificial pacemaker *in situ*, shown by X-ray

The defibrillator

The result of a heart attack or of severe shock may be that the heart muscles fail to function in sequence. The muscles function independently, leading to a failure of the heart to pump blood. The condition is called **fibrillation** and is observed on an ECG as a very irregular trace. It can lead to death in a very short time. To try to stop fibrillation, a current of about 20 A is passed through the heart for a few milliseconds using a defibrillator, as shown in Figure 3.9.

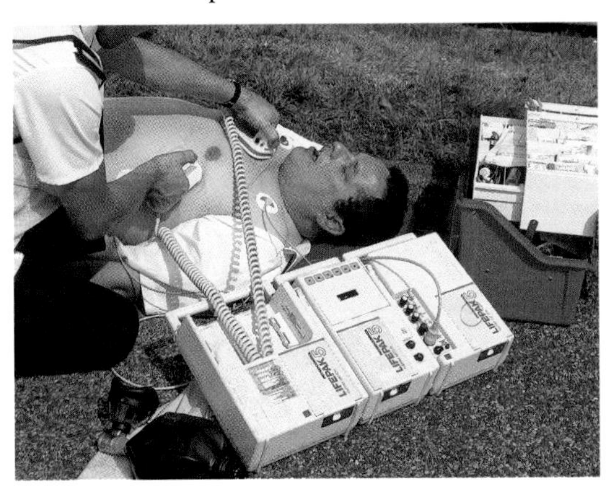

Figure 3.9 A defibrillator being used on a heart attack victim

Two electrodes, or 'paddles', are placed on the skin, one on each side of the heart. The potential difference across the electrodes is about 3000 V and the energy dissipated is of the order of 100 J. Needless to say, the operator must be insulated from the defibrillator!

The effect of this electric shock is to make all the heart muscles suffer a major contraction. This should then jolt the heart back into its normal rhythm.

Blood pressure

In order to make any fluid flow through a tube, there must be pressure differences along the length of the tube. Similarly, in order to make blood flow along arteries and veins, the blood must be at pressure. This pressure, generally measured in millimetres of mercury (mmHg), is the excess pressure in the blood above atmospheric pressure.

Since 1 atmosphere pressure corresponds to 76 cm of mercury, and, for a liquid,

$$\text{pressure} = \text{depth} \times \text{density} \times \text{gravitational field strength}$$

then

$$1.0 \text{ atmosphere} = 0.76\,\text{m} \times 13\,600\,\text{kg}\,\text{m}^{-3} \times 9.8\,\text{N}\,\text{kg}^{-1}$$

$$\approx 1.0 \times 10^5\,\text{N}\,\text{m}^{-2} \text{ or Pa (pascal)}$$

(The density of mercury is $13\,600\,\text{kg}\,\text{m}^{-3}$ and the gravitational field strength is $9.8\,\text{N}\,\text{kg}^{-1}$.)

Blood pressure of 120 mmHg corresponds to an excess pressure of approximately 1.6×10^4 Pa above atmospheric.

Table 3.1 gives typical values for blood pressure and mean blood flow rate in some blood vessels.

Table 3.1 Blood pressures and flow rates

Vessel	**Pressure** /mmHg	**Mean flow rate** /mm s^{-1}
Aorta	60–150	400
Terminal arteries	35–50	<100
Capillaries	25–10	<1
Terminal veins	<8	1
Vena cava	~2	200

It is interesting to note that, for a healthy adult, blood flows through the aorta (the artery from the heart to the body) at a rate of 3.5 to 5 litres per minute!

Detailed information on the flow profile of blood in arteries and veins can be obtained using Doppler ultrasound techniques (see Chapter 9, pages 138–140). Remember that blood vessels are elastic (some people can see their pulse at the wrist as a small artery expands and contracts) and so the comparison of blood flow in an artery with water flowing down a pipe is not very realistic.

The pumping action of the heart causes a regular change in blood pressure in the arteries. This can be felt as 'the pulse' in arteries near the body surface (e.g. the neck or the inside of the wrist). As the ventricles contract, the pressure of the blood in the arteries increases. The aortic valve (see Figure 3.5) closes when the pressure in the aorta (the back pressure) becomes greater than the pressure in the heart. The blood then flows through the arteries with a consequent decrease in pressure. A pressure wave (the pulse) passes through the arteries, gradually reducing in amplitude. This cycle, illustrated in Figure 3.10, then repeats itself.

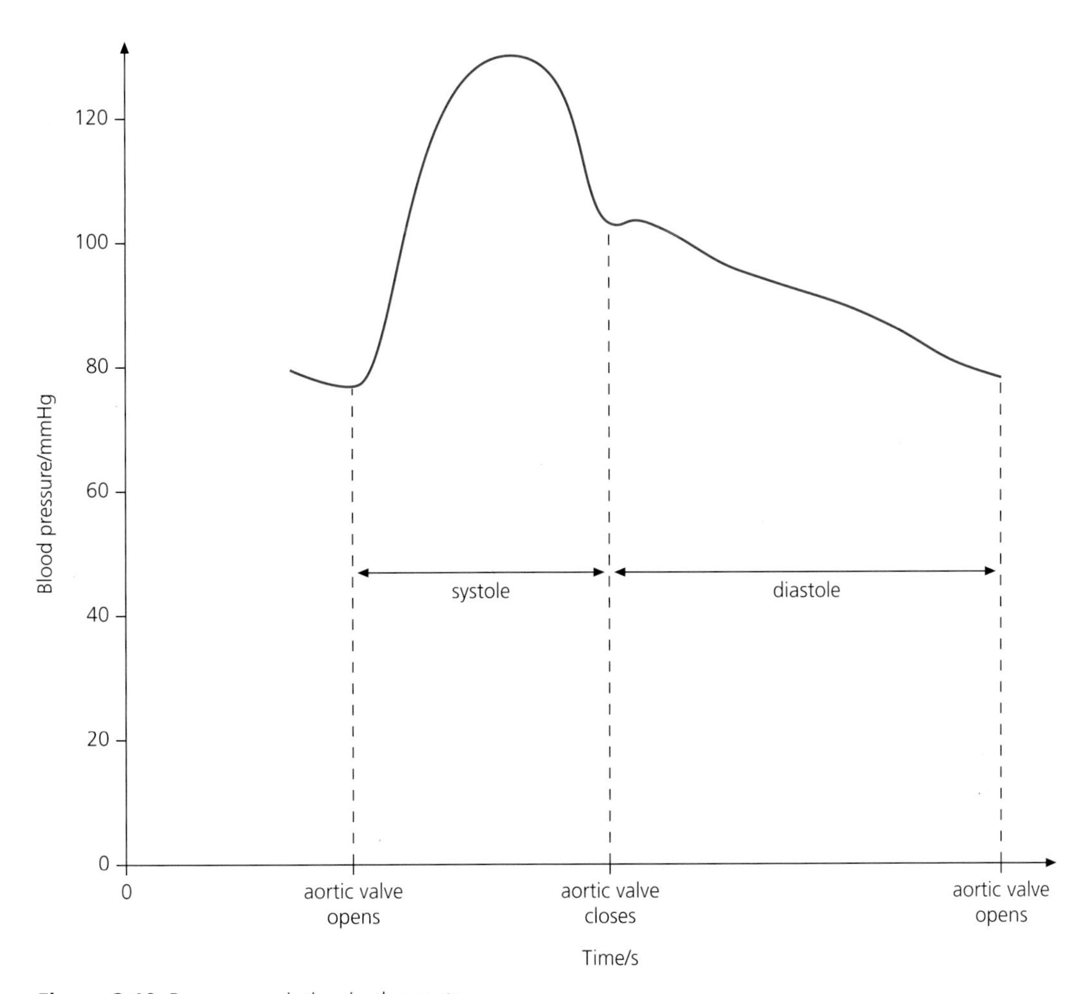

Figure 3.10 Pressure variation in the aorta

The part of the cycle when the pressure rises (between the opening and closing of the aortic valve) is called the **systole**. This is then followed by the **diastole**.

Note that the pumping of the heart and the elasticity of the blood vessels is responsible for the flow of blood through the arteries to the capillaries, whereas muscular contraction is greatly responsible for the return of blood to the heart along the veins. As the muscles contract, they squeeze the veins, forcing the blood along them. Veins have valves at regular intervals so that blood in them will flow in one direction only. Since muscular activity is important for blood flow in veins, it is for this reason that blood may become stagnant in veins if the person sits or lies still for long periods of time. The stagnant blood may then clot, causing deep vein thrombosis (DVT). If the clot subsequently moves to the heart or lungs, the results may be fatal.

The sphygmomanometer

The pressures and the pressure changes of the blood in the arteries give an indication of the health of the heart and the blood vessels. For example, high blood pressure may well lead to heart failure or the rupture of small blood vessels in the brain (a 'stroke'). Routine blood pressure measurements are made using a **sphygmomanometer**, as shown in Figure 3.11.

The instrument consists of an inflatable cuff and a mercury manometer. The cuff is placed around the arm at heart-height. It is then inflated by pumping in air and the manometer measures the pressure of this air. The pressure is increased until the blood flow in the brachial artery (in the arm) is stopped. This is detected by placing a stethoscope on the artery below the cuff (near the inside of the elbow). No sound is heard. As the pressure in the cuff is then reduced, there comes a point when blood can just spurt past the cuff. This occurs when the blood is at its maximum pressure. The turbulence produced by the spurting blood can be heard using the stethoscope. This pressure is recorded and is known as the **systolic pressure**. As the pressure in the cuff falls further, turbulence will still be heard until the brachial artery is no longer restricted. This occurs when the pressure in the artery and the cuff are the same. The pressure recorded just as the sound of turbulence is no longer heard is referred to as the **diastolic pressure**.

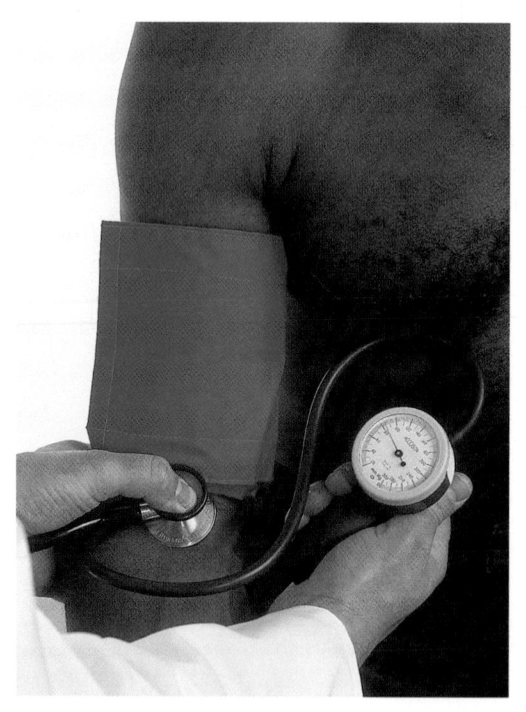

Figure 3.11 A patient's blood pressure being recorded by a sphygmomanometer

The measurement of 'the blood pressure' would be recorded as, for example, 130/80. The figures are the pressures in millimetres of mercury. The numbers 130 and 80 refer to the systolic pressure and the diastolic pressure respectively.

The sphygmomanometer is not a particularly accurate instrument and it does not indicate the pressure profile of the pulse. However, it is a simple procedure for the measurement of blood pressure and does not require any direct contact with the blood. With the development of solid state electronics, sphygmomanometers that do not use a mercury manometer have been developed, but the same pressures are measured.

Questions

1 An electrical signal passes along a nerve fibre at a speed of about $150\,m\,s^{-1}$. Estimate the time taken for a signal to travel from the base of the spine along a nerve fibre to the foot.

2 a Explain what is meant by a *resting potential* and an *action potential*.
b Describe briefly the migration of ions through the membrane of an axon giving rise to the resting and action potentials.

3 a With the aid of a labelled diagram, explain the pumping action of the human heart.
b Figure 3.12 shows part of an ECG for a particular patient.

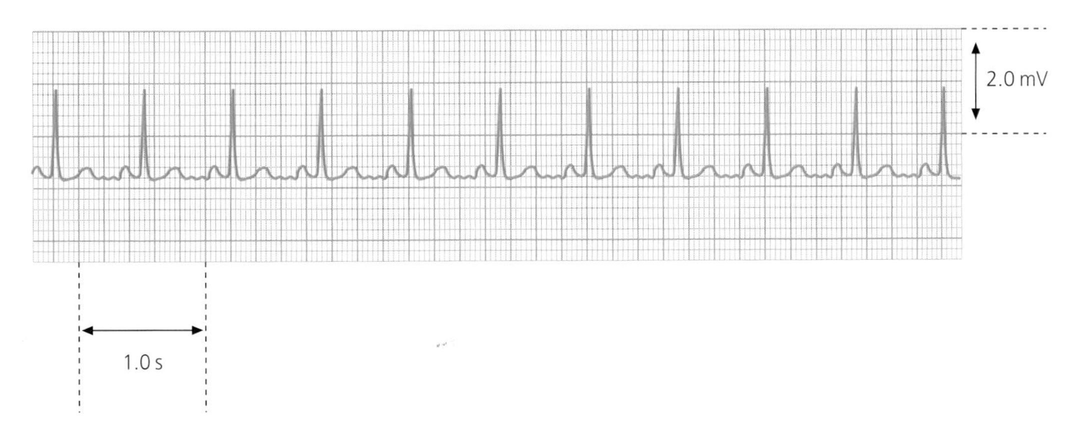

Figure 3.12

Use the trace to estimate the pulse rate of the patient.

4 a Figure 3.13 shows part of the trace produced by an electrocardiograph for a healthy person.
State the actions giving rise to sections A, B and C of this trace.

b State what change, if any, would occur in the shape of the trace in Figure 3.13 for a patient who has
i) damaged heart muscle,
ii) a partial blockage in one part of the heart.

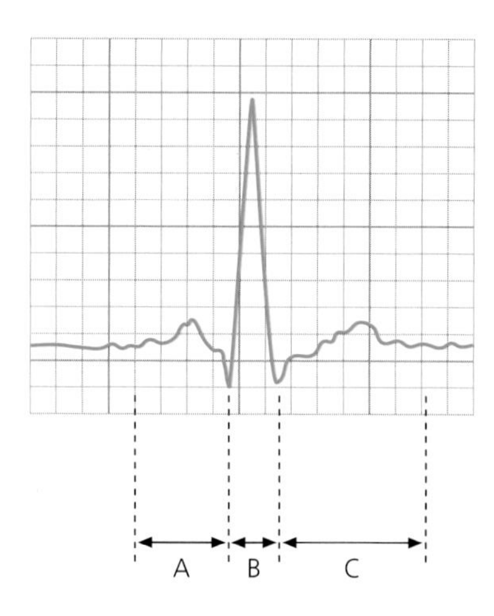

Figure 3.13

5 a Explain what is meant by the *systolic* and the *diastolic* blood pressure.

b A patient with high blood pressure is told that the pressure is 165/100.
i) Explain what is meant by '165/100'.
ii) Calculate the maximum pressure recorded. The density of mercury is $13600\,\mathrm{kg\,m^{-3}}$ and the gravitational field strength is $9.8\,\mathrm{N\,kg^{-1}}$.

4 Body mechanics

In this chapter you will read about:

- how bones, joints, muscles, ligaments and tendons produce movement
- how combinations of muscles and bones act to produce lever effects
- the forces involved in bending and lifting
- the forces involved in walking and running
- the energy requirements of the human body

Figure 4.1

The structure of the human body

The human body is made up of over 200 bones arranged to support the body, protect vital organs and allow movement. Where two of these bones meet, a **joint** is formed (see Figures 4.2 and 4.3 overleaf).

Some joints allow movement in one plane only, such as that found in the knee, shown in Figure 4.2. This type of joint is called a **hinge**. Other joints allow movement in more than one plane, such as the joint between the femur and the pelvis, shown in Figure 4.3. This type of joint is called a **ball and socket** joint.

Some joints do not allow any movement at all, for example the joints between the bones that make up the skull.

The bones at a joint are moved by muscles. These are attached to the bones by **tendons**. Muscles are only able to effect a force when they contract. Fibrous tissues called **ligaments** hold the bones in place within the joint during muscular movement (see Figure 4.3).

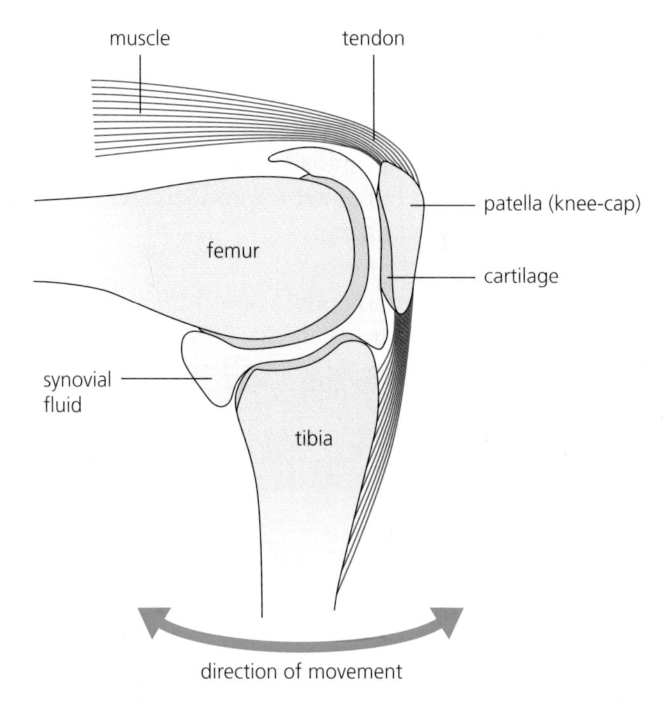

Figure 4.2 Hinge joint in the knee

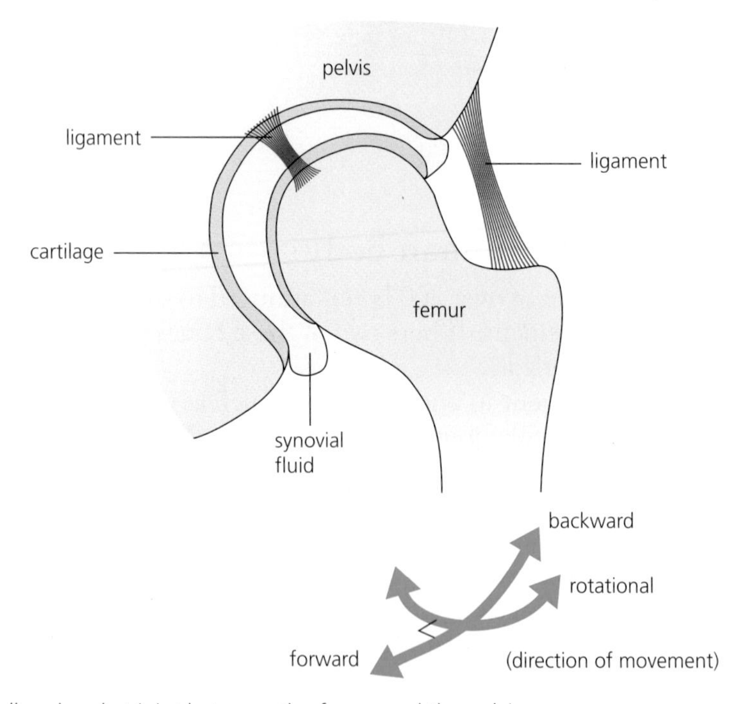

Figure 4.3 Ball and socket joint between the femur and the pelvis

Lever systems in the body

A lever system allows an input force called the **effort** to move (or balance) an output force called the **load**. The effort causes a turning effect about a pivot that acts in an opposing sense to the turning effect of the load. This turning effect is called a **moment** and is defined as the product of the force and the perpendicular distance of the line of action of this force from the pivot (or fulcrum):

$$\text{moment} = \text{force} \times \text{perpendicular distance (from the pivot)}$$

If the lever system is stationary, the turning forces acting on it are balanced and the lever system is said to be in **equilibrium**. An example of this is shown in Figure 4.4.

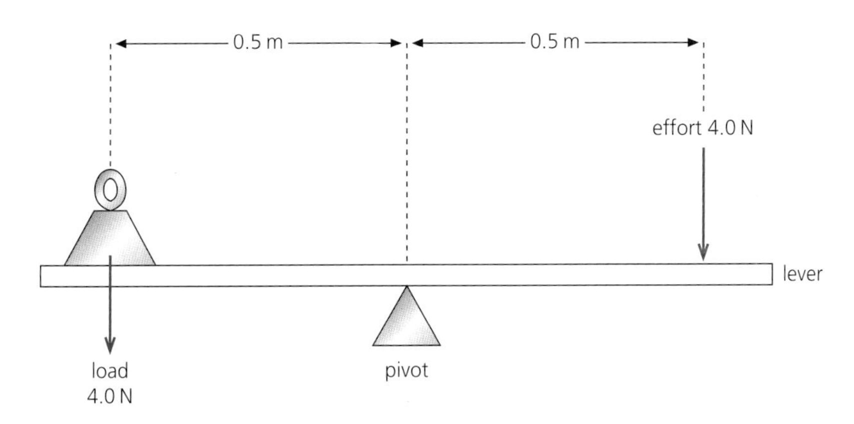

Figure 4.4 A balanced lever

In this situation, the clockwise moment of the effort about the pivot:

$$4.0\,\text{N} \times 0.50\,\text{m} = 2.0\,\text{N m}$$

is balanced by the anticlockwise moment of the load about the pivot:

$$4.0\,\text{N} \times 0.50\,\text{m} = 2.0\,\text{N m}$$

The lever is uniform so its centre of mass acts mid-way along its length. As this centre of mass is above the pivot, there is no moment about the pivot due to the weight of the lever.

When the effort is applied at a greater distance from the pivot than that of the load, the lever system acts as a **force multiplier**. This means the effort necessary to balance the load is less. Figure 4.5 shows such a situation.

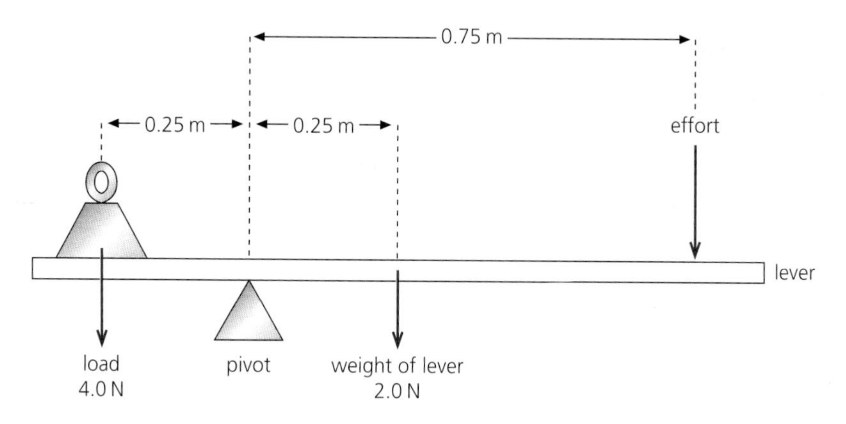

Figure 4.5 A lever acting as a force multiplier

The 4.0 N load now causes an anticlockwise moment of:

$$4.0\,\text{N} \times 0.25\,\text{m} = 1.0\,\text{N m}$$

As the pivot is no longer directly beneath the centre of mass of the lever, the 2.0 N weight of the lever now has a net clockwise moment about the pivot of:

$$2.0\,\text{N} \times 0.25\,\text{m} = 0.50\,\text{N m}$$

For the lever to remain horizontally in equilibrium, the effort must cause the remaining clockwise moment of 0.50 N m. This moment is divided by 0.75 m, the perpendicular distance from the pivot of the line of action of the effort, giving a value for the effort of only 0.67 N.

It is useful to describe the relationship between the effort required to move a load and the magnitude of the load. This is achieved by quoting the **mechanical advantage** of the lever system. The mechanical advantage of a lever system is given by either:

$$\text{mechanical advantage} = \frac{\text{load}}{\text{effort}}$$

or, if the weight of the lever is negligible:

$$\text{mechanical advantage} = \frac{\text{perpendicular distance of the effort from the pivot}}{\text{perpendicular distance of the load from the pivot}}$$

If the output force is bigger than the input effort, the system is described as having a mechanical advantage of 'greater than 1'. A mechanical advantage of 'less than 1' means that, to enable the system to be in equilibrium, the effort applied must be greater than the load. Where bones and muscles in the body act as levers the mechanical advantage is usually less than 1. In other words, the muscles are contracting and applying a force to the bones that is greater than the load being moved. This is advantageous as it results in a smaller movement of the muscles.

WORKED EXAMPLE 4.1

Figure 4.6 shows a lower arm of weight 50 N. The arm supports a weight of 20 N. The perpendicular distances from the pivot to the lines of action of the 50 N weight and the 20 N weight are 0.30 m and 0.65 m respectively. The arm maintains this position as the biceps muscle pulls on the bones of the lower arm at a distance from the pivot of 0.030 m. Calculate:

a the force needed by the biceps muscle to maintain the lower arm in equilibrium,
b the mechanical advantage of the arm when maintaining this position (ignoring the weight of the lower arm).

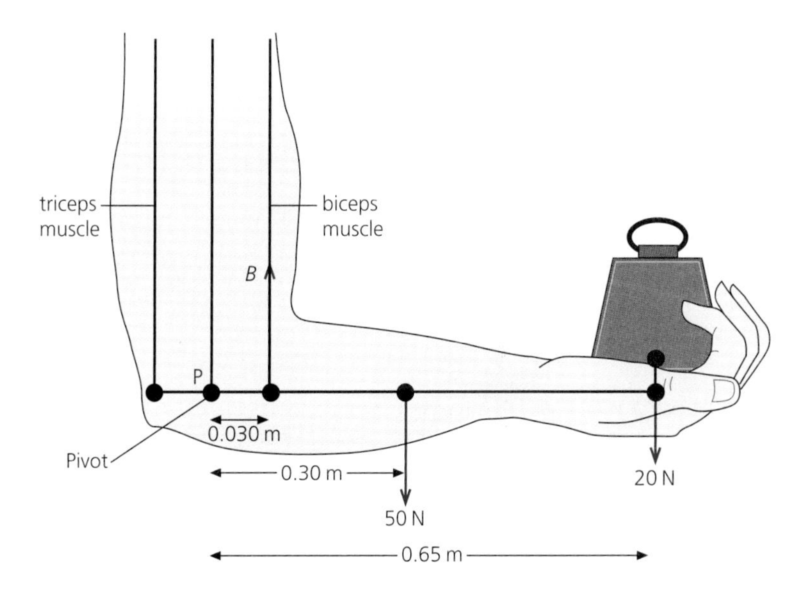

Figure 4.6

a For equilibrium, the sum of the clockwise moments about the pivot P is equal to the sum of the anticlockwise moments. Let the force of the biceps be B.

$$(50 \times 0.30)\,\mathrm{N\,m} + (20 \times 0.65)\,\mathrm{N\,m} = B \times 0.030\,\mathrm{m}$$

$$(15 + 13)\,\mathrm{N\,m} = B \times 0.030\,\mathrm{m}$$

$$B = \frac{28\,\mathrm{N\,m}}{0.030\,\mathrm{m}} = 933\,\mathrm{N}$$

b Mechanical advantage $= \dfrac{\text{load}}{\text{effort}}$

$$= \frac{20\,\mathrm{N}}{933\,\mathrm{N}}$$

$$= 0.021$$

Bending and lifting

When a human body bends at the waist, large forces are exerted by the muscles in the back. In order to lift a load from the floor, a body may bend forwards at the waist, keeping the legs relatively straight (Figure 4.7a). The centre of mass of the upper body (C) causes a clockwise moment about the pivot – the lumbosacral joint (J). If the back aligns roughly in a horizontal position then the perpendicular distance from the centre of mass to the pivot is a maximum, as is the distance from the pivot of the load to be lifted. This causes a maximum clockwise moment that must be counterbalanced by the back muscles if the stance is to be maintained. As the position of the back muscles is fixed, the only way to increase the anticlockwise moment is to increase the force applied by the muscles. This puts a great strain on the tendons in the back and increases the potential for damage.

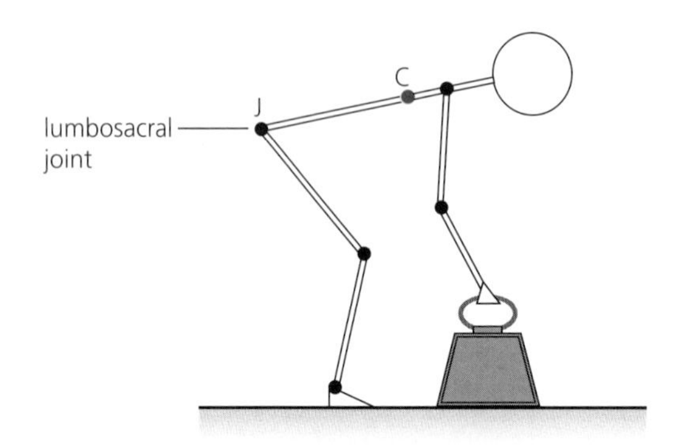

(a) This way of bending and lifting puts great strain on the back muscles

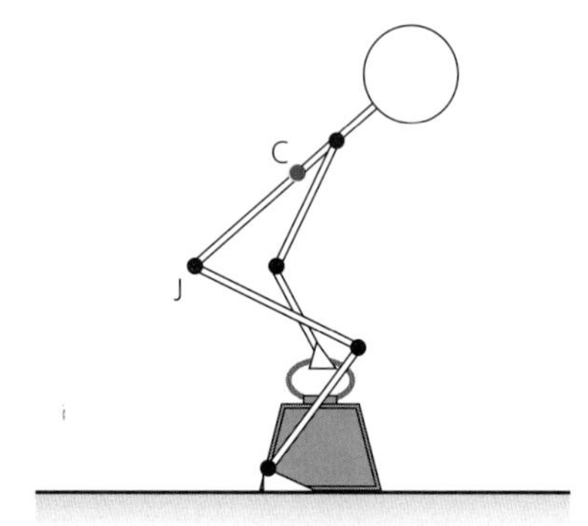

(b) This way reduces the strain

Figure 4.7

Figure 4.8 Lifting a load correctly

In order to reduce the strain on back muscles during bending and lifting it is necessary to reduce both the perpendicular distance of the centre of mass of the back from the pivot, as well as the distance from the pivot of the load to be lifted. This is achieved by keeping the line of the backbone as close to the vertical as possible, bending the knees and moving into a position so that the load to be lifted is directly below the centre of mass of the person (Figures 4.7b and 4.8). The magnitude of the moments is greatly reduced and the lifting forces are now being supplemented by the much more powerful leg muscles.

WORKED EXAMPLE 4.2

Figure 4.9 shows a schematic diagram of the upper body of a person who is bending forwards so that the line of the back is inclined at an angle of 60° to the vertical.

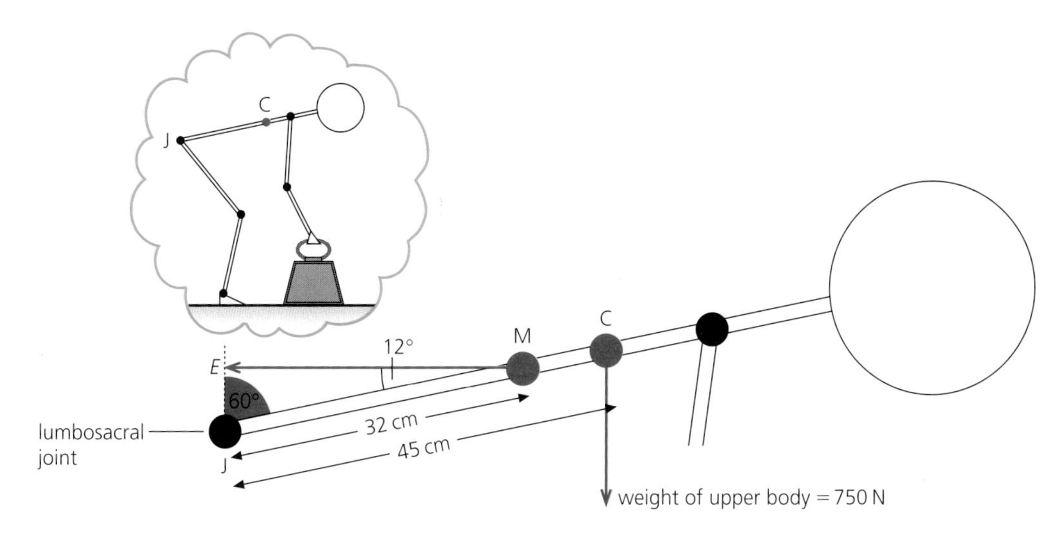

Figure 4.9

The upper body moves about a point called the lumbosacral joint (J). The weight of the upper body is 750 N. This weight acts through a point C, the centre of mass of the upper body, which is situated at a distance 45 cm from J. The back muscles may be considered to cause a net single anticlockwise moment about the lumbosacral joint. The force *E*, exerted by the back muscles, acts along a line through a point M on the body and inclined at an angle of 12° to the line JMC. Point M is situated at a distance of 32 cm from J.

a Calculate the clockwise moment due to the weight of the body about the lumbosacral joint.

b Give an equation for the anticlockwise moment of the force *E* due to the back muscles about the lumbosacral joint (in terms of *E*).

c When the body holds the stance shown in Figure 4.9, the body is considered to be in equilibrium and the sum of the clockwise moments is equal to the sum of the anticlockwise moments. By equating the answers to **a** and **b**, calculate a value for the force *E*, due to the back muscles, needed to sustain the position shown.

d When bending in order to lift a load, the muscles in the back act not only to raise the load but also to support the upper body. Explain, with reference to Figure 4.10 overleaf, why bending the legs and keeping the line of the back as close to the vertical as possible is advisable when lifting heavy loads.

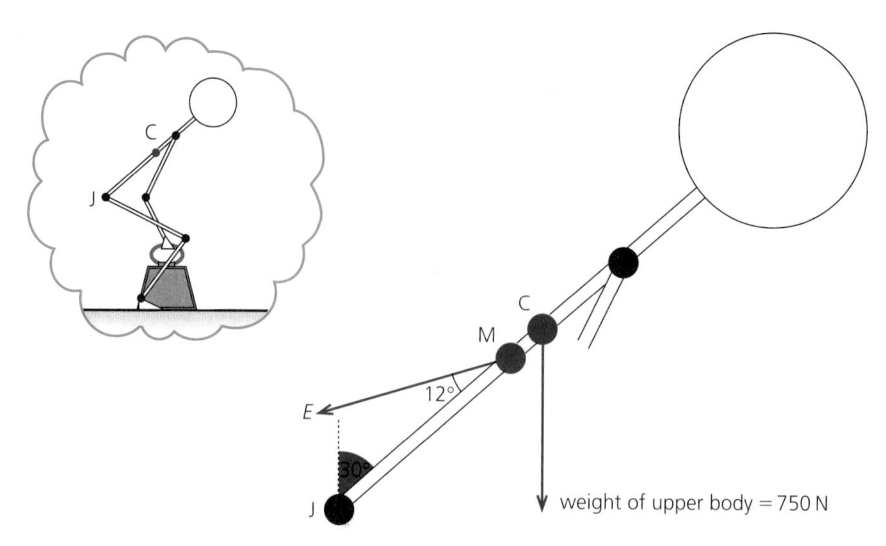

Figure 4.10

a Clockwise moment about J = force × perpendicular distance to the pivot

$$= 750\,\text{N} \times 0.45\,\text{m} \times \sin 60° \quad \text{(see Figure 4.11a)}$$
$$= 292\,\text{N m}$$

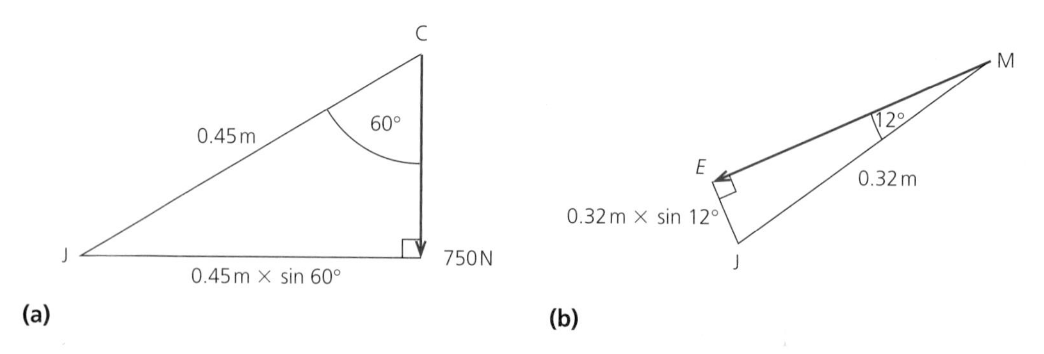

(a) (b)

Figure 4.11

b Anticlockwise moment about J = $E \times 0.32\,\text{m} \times \sin 12°$ (see Figure 4.11b)

c

$$E \times 0.32\,\text{m} \times \sin 12° = 292\,\text{N m}$$

$$E = \frac{292}{0.32 \sin 12°}\,\text{N}$$

$$= 4400\,\text{N}$$

d The clockwise moment of the weight of the body about the lumbosacral joint is directly proportional to the sine of the angle which the back makes with the vertical.

In Figure 4.10, assuming the same body weight and dimensions as in Figure 4.9,

$$\text{clockwise moment} = 750\,\text{N} \times 0.45\,\text{m} \times \sin 30° = 169\,\text{N m}$$

(Note this is less than the 292 N m in answer **a**.) So as the angle that the back makes with the vertical becomes less, so the clockwise moment becomes less and the force *E* from the back muscles becomes less. If the back is kept nearer vertical then it becomes necessary to bend the knees in order to lift a load. There is then less strain on the back muscles while the leg muscles do more work.

The forces involved in walking and running

When a person stands at rest, the weight of the body acts vertically downwards. The ground exerts an upward force on the feet of the body, equal in magnitude to the weight (Figure 4.12).

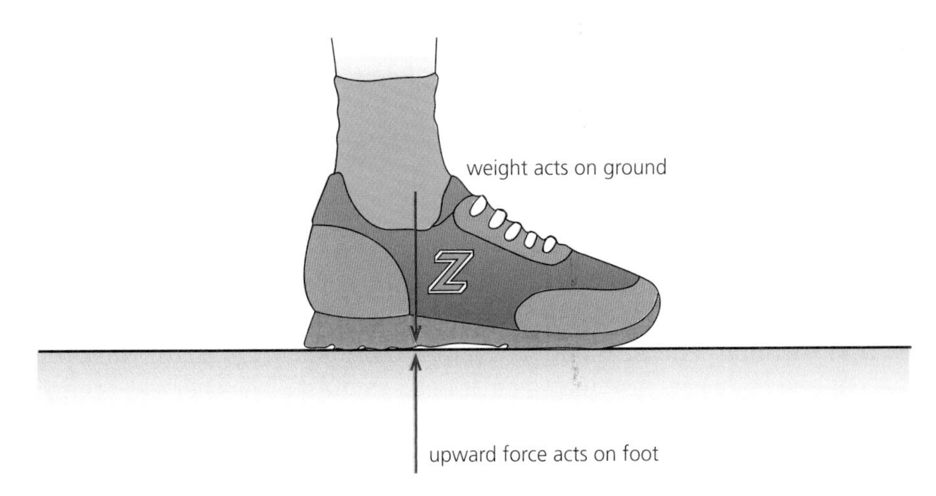

Figure 4.12 Force on the foot when standing still

When the person begins to walk forwards, the heel of one foot is raised and the sole of the foot pushes back and down. The ground pushes forward and up on this foot (see Figure 4.13a overleaf). The calf muscle contracts, pulling the heel up and giving the leg a turning effect about the ball of the foot that causes the leg to accelerate forwards and upwards.

Meanwhile the other foot, which has been lifted by the leg and swung forward about the hip joint, makes contact with the ground. This foot provides both a forward force on the ground due to the rate of change of the momentum of the body and a downward force due to the weight of the body. The ground exerts a frictional force and a reaction force on the foot, which together act for a time to decrease the momentum of the person and hence decelerate the foot (Figure 4.13b).

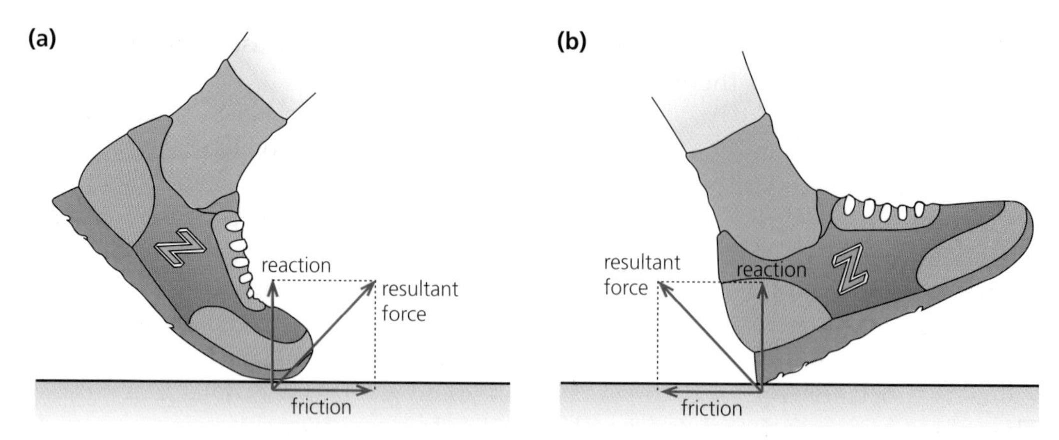

Figure 4.13 Forces on the foot when walking

When a person runs, the knee bends and straightens as the foot pushes forward. This has the effect of lifting the centre of mass of the body, giving the body an upward and forward momentum. The whole body leaves the ground until the other foot makes contact with the ground. The leg bends as the foot touches the ground, so reducing the decelerating force. It then straightens on take-off, so once again increasing the accelerating force.

The body's energy requirements

When muscles move within a body, work is done by the body. The efficiency of transfer of the chemical energy in food into mechanical energy in the muscles is typically between 10 and 30% depending on the muscles involved. The remaining 70 to 90% of the input energy is converted into thermal energy in the muscles. This explains why the body gets hot during prolonged exercise. The thermal energy produced is transferred from the body by the mechanisms of conduction, convection, radiation and evaporation.

1 Thermal energy is conducted away from the body by the medium in contact with the skin.
2 If this medium is moving with respect to the body, energy is also transferred by forced convection. This situation occurs when thermal energy is transferred from the skin of a person to the surrounding air which is then moved due to a gust of wind.
3 As the body is at a temperature higher than that of the surroundings, there is net emission of infrared radiation from the body to the surroundings.
4 When the body perspires (sweats), thermal energy (**latent heat**) is required for the change of state of the sweat from liquid to vapour. This energy is taken from the surface molecules within the sweat, lowering the average thermal energy within the sweat and hence its temperature. The body transfers heat at a greater rate to the cold sweat than directly to the surrounding air, increasing the cooling effect.

When the body is cold, it involuntarily shivers, automatically using muscular movement that produces thermal energy.

The energy required by the body for movement is supplied by the food eaten and is mostly stored as fat. The energy converted each second by a body to maintain essential bodily processes is called the **basal metabolic rate** (BMR). The value of the BMR for an average adult is about 80 W. This energy is used to keep the body alive by fuelling the muscles of the heart and lungs, enabling respiration and maintaining normal body temperature. The body converts energy at the basal metabolic rate during periods of sleep. All positions other than lying horizontally require muscular movement and hence extra energy conversion each second. Table 4.1 shows typical values of the rate of conversion of energy of an average body for various activities. The values include the 80 W required for basal metabolic processes.

Table 4.1 The body's rate of energy conversion during different activities

Activity	Rate of energy conversion/W
Sleeping	80
Sitting	120
Walking	250
Jogging	400
Swimming	500

An active person converts energy at a greater rate than compared with an inactive person. This results in the requirement of a greater input of energy from food.

WORKED EXAMPLE 4.3

Use Table 4.1 to give an estimate of the total energy requirement over a 24-hour period of:

a an adult who spends 8 h sleeping, 8 h sitting, working in an office, and 8 h sitting, watching television,

b an adult who spends 9 h sleeping, 7 h walking at work, 1 h jogging, 2 h walking to and from work, 1 h swimming, and 4 h sitting and relaxing in the evening.

a Sleeping: time spent $= 8\,\text{h} = 8 \times 60 \times 60\,\text{s} = 28\,800\,\text{s}$
Rate of energy conversion $= 80\,\text{W} = 80\,\text{J}\,\text{s}^{-1}$
So energy expended $= 80\,\text{J}\,\text{s}^{-1} \times 28\,800\,\text{s} = 2.3 \times 10^6\,\text{J}$

Sitting: time spent at work and at home $= 16 \times 60 \times 60\,\text{s} = 57\,600\,\text{s}$
Rate of energy conversion $= 120\,\text{W} = 120\,\text{J}\,\text{s}^{-1}$
So energy expended $= 120\,\text{J}\,\text{s}^{-1} \times 57\,600\,\text{s} = 6.9 \times 10^6\,\text{J}$

Total energy expended $= (2.3 \times 10^6 + 6.9 \times 10^6)\,\text{J} = 9.2 \times 10^6\,\text{J}$

b Sleeping: time spent $= 9 \times 60 \times 60\,\mathrm{s} = 32\,400\,\mathrm{s}$
Rate of energy conversion $= 80\,\mathrm{W} = 80\,\mathrm{J\,s^{-1}}$
So energy expended $= 80\,\mathrm{J\,s^{-1}} \times 32\,400\,\mathrm{s} = 2.6 \times 10^6\,\mathrm{J}$

Walking: time spent $= (7 + 2) \times 60 \times 60\,\mathrm{s} = 32\,400\,\mathrm{s}$
Rate of energy conversion $= 250\,\mathrm{W} = 250\,\mathrm{J\,s^{-1}}$
So energy expended $= 250\,\mathrm{J\,s^{-1}} \times 32\,400\,\mathrm{s} = 8.1 \times 10^6\,\mathrm{J}$

Jogging: time spent $= 1 \times 60 \times 60\,\mathrm{s} = 3600\,\mathrm{s}$
Rate of energy conversion $= 400\,\mathrm{W} = 400\,\mathrm{J\,s^{-1}}$
So expended energy $= 400\,\mathrm{J\,s^{-1}} \times 3600\,\mathrm{s} = 1.4 \times 10^6\,\mathrm{J}$

Swimming: time spent $= 1 \times 60 \times 60\,\mathrm{s} = 3600\,\mathrm{s}$
Rate of energy conversion $= 500\,\mathrm{W} = 500\,\mathrm{J\,s^{-1}}$
So expended energy $= 500\,\mathrm{J\,s^{-1}} \times 3600\,\mathrm{s} = 1.8 \times 10^6\,\mathrm{J}$

Sitting: time spent $= 4 \times 60 \times 60\,\mathrm{s} = 14\,400\,\mathrm{s}$
Rate of energy conversion $= 120\,\mathrm{W} = 120\,\mathrm{J\,s^{-1}}$
So energy expended $= 120\,\mathrm{J\,s^{-1}} \times 14\,400\,\mathrm{s} = 1.7 \times 10^6\,\mathrm{J}$

Total energy expended $= (2.6 \times 10^6 + 8.1 \times 10^6 + 1.4 \times 10^6 + 1.8 \times 10^6 + 1.7 \times 10^6)\,\mathrm{J}$
$= 1.6 \times 10^7\,\mathrm{J}$

The more active a person is, the greater the energy requirements of that person.

Questions

1 Figure 4.14 shows an arm supporting a weight of 15 N at a distance of 35 cm from the elbow joint J. The lower arm has a weight of 25 N which acts through a point 25 cm from J. The biceps muscle is in tension and acts on the bones of the lower arm at a distance of 3.0 cm from J.

a Calculate the tension *B* in the biceps muscle.

b Calculate the mechanical advantage of the arm when supporting the weight.

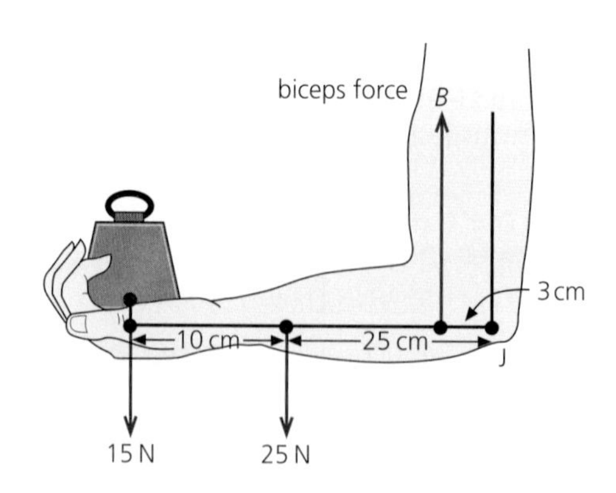

Figure 4.14

2 Explain with reference to Figure 4.15 which stance, **A** or **B**, should be adopted when lifting a heavy load from the floor onto the back of a lorry.

Figure 4.15

3 Describe the forces that act between the foot and the ground for a person who is walking,

- **a** as the front foot strikes the ground,
- **b** as the centre of mass of the person is directly above the foot,
- **c** as the foot leaves the ground.

4 The basal metabolic rate of an average adult is 80 W.

- **a** Explain the meaning of the term *basal metabolic rate.*
- **b** Calculate the minimum energy required when sleeping for a period of 8 hours if the metabolic rate for this activity is 80 W.
- **c** In practice, the magnitude of the energy requirement of a person who sleeps for 8 hours differs from that calculated in **b**. Explain why.
- **d** The metabolic rate for a person who is walking along a level road is 250 W. Explain the conversions of energy within the body that lead to this rate of 250 W.

5 Figure 4.16 shows a head of weight 150 N supported by a bone called the atlas vertebra. The neck muscle acts at a distance of 0.040 m from the atlas vertebra while the centre of mass of the head acts along a line 0.025 m the other side of the atlas vertebra.

Calculate the effort in the neck muscle required to maintain the head in equilibrium.

0.025 m
0.040 m
150 N
neck muscle
(effort)
atlas
vertebra

Figure 4.16

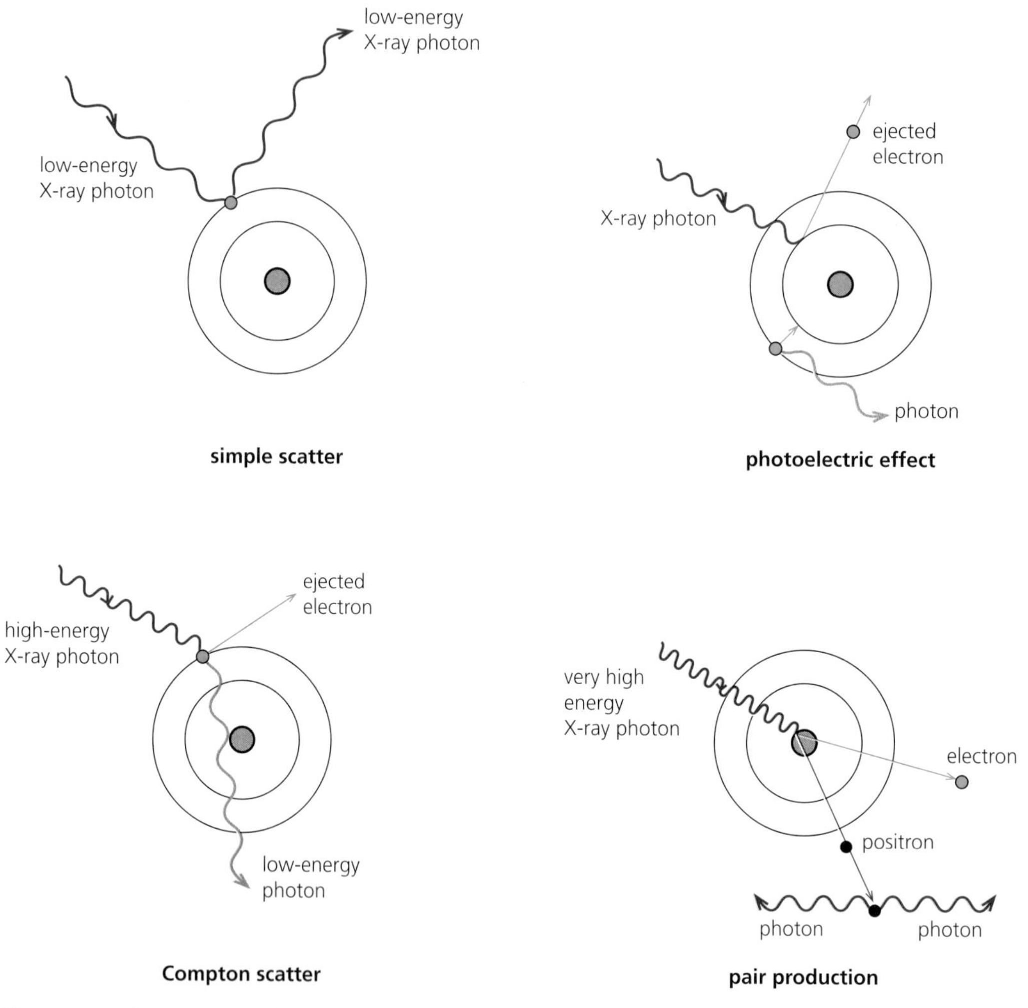

Figure 5.2 The processes of attenuation

1 Simple scatter

The energy of the photon is less than the energy required to remove an electron from its atom (i.e. to cause ionisation, see 2 below). The photon changes direction. The power of the beam in the original direction of travel is reduced.

2 Photoelectric effect

The photon has an energy greater than the binding energy of an electron in the atom. The photon is annihilated and its energy is transformed to eject the electron from the atom. This is **ionisation**. Another electron in the atom then moves from a higher energy state to fill the vacancy and, during this process, a lower energy photon is emitted.

3 Compton scatter

The photon energy far exceeds the binding energy of the electrons in the atom. An electron is ejected from the atom and a lower energy photon moves off. The direction of the photon is different from that of the incident photon in order to conserve momentum. Note that, in the photoelectric effect, the photon is completely annihilated: all its energy is given up to the ejected electron. In Compton scatter, a lower energy photon results. The lower energy photon may then give rise to further Compton scatter.

4 Pair production

This occurs only at very high photon energies, in excess of about 1 MeV. The photon interacts with the nucleus of the atom. The photon is annihilated, creating an electron and a positron. A positron is a particle having the same mass as an electron and the same magnitude of charge but with opposite sign (positive). The two particles lose energy by ionisation as they move apart. The positron is subsequently annihilated by an electron, creating two gamma ray photons, each of energy 0.51 MeV.

Attenuation coefficients

When electromagnetic radiation is radiated uniformly in all directions from a point source in a vacuum, the radiation energy is not absorbed. However, the intensity is reduced with distance – that is, the beam is attenuated. The energy is spread out over the area of a sphere with its centre at the source of the radiation (see Figure 5.3).

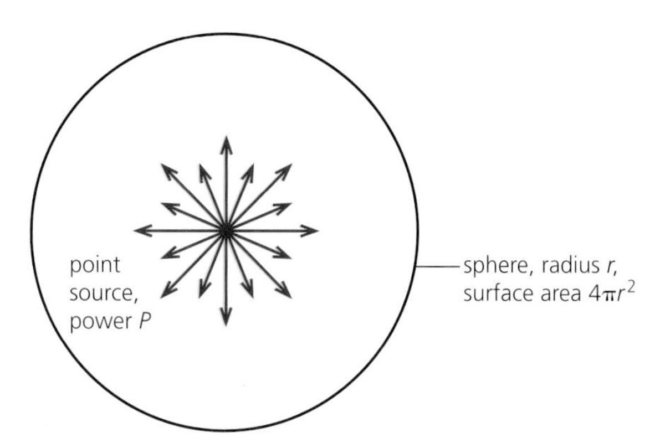

Figure 5.3 Attenuation with distance

If the power of the source is P, then at distance r from the source, the intensity I is given by:

$$I = \frac{P}{4\pi r^2}$$

This means that one of the simplest and yet effective means of preventing excessive exposure to radiation is to keep away!

In any medium, energy **absorption** processes take place and the radiation will be further attenuated. The intensity I of a parallel beam of radiation having passed through a thickness x of the medium is given by:

$$I = I_0 e^{-\mu x}$$

where I_0 is the incident intensity (the intensity for $x = 0$) and μ is a constant called the **linear attenuation coefficient** (or sometimes the **linear absorption coefficient**). μ is a constant dependent on the medium and on the photon energy of the radiation. Its unit is metre^{-1} (m^{-1}), where x is measured in metres. The variation with thickness x of the intensity I is illustrated in Figure 5.4.

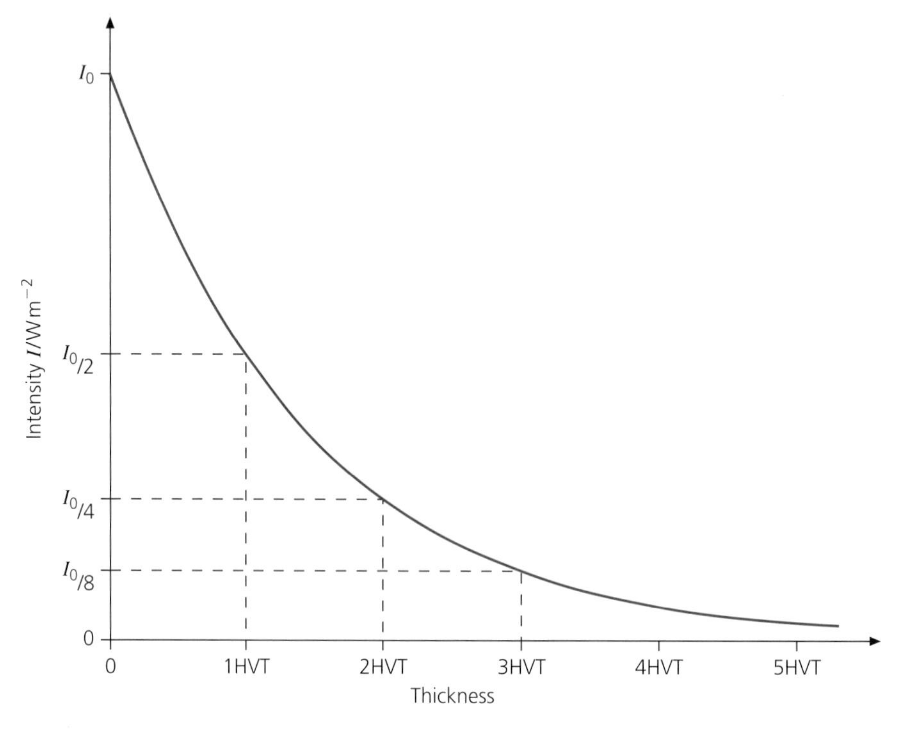

Figure 5.4 Exponential attenuation due to absorption

The graph is exponential and thus the thickness of the medium required to completely 'stop' the radiation cannot be determined. Instead, the penetrating power, or **quality** of the radiation, is described in terms of the thickness of the medium required to reduce the intensity to one half of its initial value. This thickness is referred to as the **half-value thickness** (HVT) and is a constant for a particular medium, as the graph shows. It is given the symbol $x_{\frac{1}{2}}$.

Since $I = I_0 e^{-\mu x}$ and $I = \frac{1}{2} I_0$ when $x = x_{\frac{1}{2}}$,

$$\tfrac{1}{2} = e^{-\mu x_{\frac{1}{2}}}$$

and

$$2 = e^{\mu x_{\frac{1}{2}}}$$

Taking logarithms to the base e,

$$\ln 2 = \mu x_{\frac{1}{2}} \quad (\text{where } \ln = \log_e)$$

$$x_{\frac{1}{2}} = \frac{\ln 2}{\mu} = \frac{0.693}{\mu}$$

An alternative to the linear attenuation (absorption) coefficient for expressing absorption of radiation is the **mass attenuation coefficient** or **mass absorption coefficient**, denoted by μ_m. This is determined by dividing the linear attenuation coefficient μ by the density ρ of the absorbing medium. Thus,

$$\mu_m = \frac{\mu}{\rho}$$

The unit of μ_m is $m^2\,kg^{-1}$ and it gives a measure of the effective mass of absorber per unit area presented to the radiation.

WORKED EXAMPLE 5.1

A parallel beam of 80 keV X-ray photons has a half-value thickness (HVT) in copper of 1.0 mm. The density of copper is $8900\,kg\,m^{-3}$. Calculate, for this beam in copper,

a the linear attenuation coefficient,
b the mass attenuation coefficient,
c the fraction of the incident intensity of radiation remaining after a thickness of 2.5 mm of copper.

a $I = I_0 e^{-\mu x}$

For $x = x_{\frac{1}{2}} = 1.0\,mm$, $I = \frac{1}{2}I_0$:

$$\tfrac{1}{2}I_0 = I_0 e^{-\mu \times 1.0}$$

$$2 = e^{\mu \times 1.0}$$

$$\ln 2 = \mu \times 1.0$$

$$\mu = \frac{\ln 2}{1.0} = 0.693\,mm^{-1} = 693\,m^{-1}$$

b

$$\mu_m = \frac{\mu}{\rho}$$

$$= \frac{693}{8900} = 0.078\,m^2\,kg^{-1}$$

c

$$\frac{I}{I_0} = e^{-0.693 \times 2.5} = 0.18$$

Attenuation and photon energy

The mass attenuation coefficient μ_m depends on the number of electrons present in the medium with which the photons can interact. Thus, μ_m depends on the proton number (atomic number) Z of the medium. Attenuation also depends on photon energy since the mechanisms of attenuation (see page 70) are energy dependent. The dependence of attenuation mechanisms on proton number and on photon energy are shown in Table 5.1.

Table 5.1 Attenuation mechanisms

Mechanism	Dependence on proton number Z	Dependence on photon energy E	Energies for which mechanism is dominant in soft tissue
Simple scatter	$\propto Z^2$	$\propto E^{-1}$	1–30 keV
Photoelectric effect	$\propto Z^3$	$\propto E^{-3}$	1–100 keV
Compton scatter	no dependence	gradual fall-off with increasing E	0.5–5 MeV
Pair production	$\propto Z^2$	increases slowly with increasing E	>5 MeV

The total attenuation of radiation in a medium is the sum of the contributions of the individual attenuation mechanisms. This is illustrated in Figure 5.5, where the mass attenuation coefficient has been plotted against photon energy.

It should be remembered that, as the mass attenuation coefficient *decreases*, the penetration of the radiation *increases*. Also, note that if the linear attenuation coefficient had been considered, rather than mass attenuation coefficient, the absorbing medium would have to be specified. Use of the mass attenuation coefficient removes that need.

It can be seen from Figure 5.5 that for photon energies up to about 100 keV, mass attenuation coefficient decreases rapidly with increase of photon energy. This means that the penetration of X-rays can be controlled by means of the X-ray tube voltage, since it is this voltage that determines the maximum X-ray photon energy. Also, at photon energies of about 20–80 keV (the photon energies used in radiography), the predominant attenuation mechanism is the photoelectric effect and this is dependent on Z^3 (see Table 5.1). Consequently, bones that are rich in high proton number elements (e.g. calcium, $Z = 20$) absorb X-rays far more than the soft tissues with low proton numbers that surround the bone (Z for muscle is about 7.4). This results in the familiar X-ray 'shadow photographs' of bones. The use of X-radiation for both diagnostic and therapeutic purposes will be discussed in Chapter 6 (page 84).

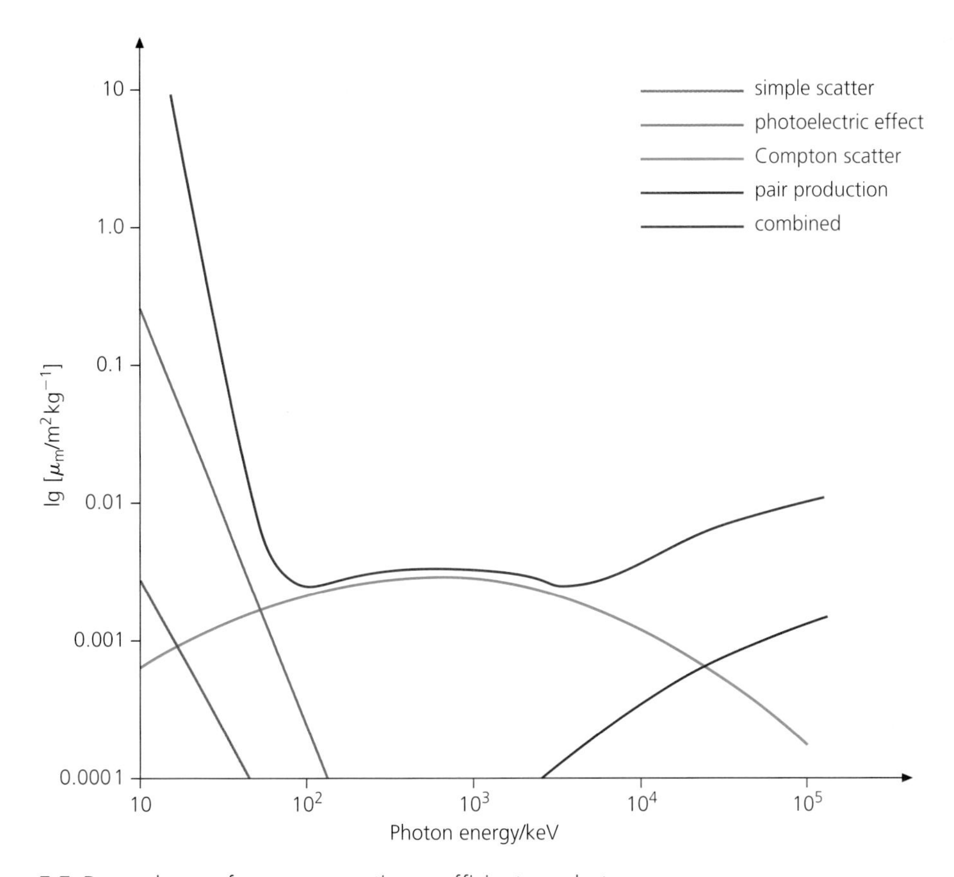

Figure 5.5 Dependence of mass attenuation coefficient on photon energy

Biological effects of radiation

When radiation is incident on a body, some is reflected (scattered), some absorbed and some transmitted. The mechanism of absorption depends on the photon (or particle) energy. For example, low photon energies may cause damage to skin tissues (for example, by too much sun-bathing) whereas excessive exposure to X-rays could produce tumours.

We saw in the previous section (page 70) that one mechanism by which radiation deposits its energy in matter is ionisation. This is the dominant mechanism for all high-energy radiations. This ionisation will take place in the cells of tissues, giving rise to damage to important molecules.

The ionisation may directly damage macromolecules such as DNA, RNA and enzymes; this is referred to as **direct damage**. Alternatively, since there is a relatively large amount of water in cells, water molecules are likely to be ionised to form OH^- and H^+ free radicals. Two OH^- ions may then react to form hydrogen peroxide (H_2O_2). This hydrogen peroxide may then damage DNA in chromosomes in the nuclei of cells. The chemical damage is referred to as **indirect damage**.

The result of the damage may be the death of the cell or its inability to reproduce or to function correctly. The overall effects of radiation are divided into two classes, **somatic** and **hereditary**. Somatic effects result from damage to cells in the body that affect the individual irradiated. On the other hand, if reproductive organs are irradiated, hereditary effects may be caused in that the person's offspring may be affected. The effects are summarised in Table 5.2.

Table 5.2 The biological effects of radiation

Damage to	Results in
Molecules	Ionisation, then chemical reaction with H_2O_2, in particular of macromolecules, e.g. DNA, RNA, enzymes
Cells	Cell death *or* failure of cell division and/or function *or* uncontrolled cell division
Organs and tissues	Disruption of functions, e.g. failure of nervous system, bone marrow function *or* destruction of organs, e.g. skin 'burns' *or* onset of malignancy, e.g. cancers
Whole body	Death within a few weeks *or* reduction in life expectancy
Populations	Changes in the 'gene pool' Life expectancy reduction Increased numbers with genetic abnormalities

Measuring radiation

The effects of radiation on living tissue depend on the quality of the radiation (for example, its ionising power) and the length of time for which the tissue is irradiated. Radiation can be harmful and, so that the level of any risk can be assessed, it is necessary to obtain quantitative measurements of radiation.

Exposure

When living tissue is irradiated, the tissue is said to have been **exposed** to the radiation. **Exposure** is a measure of the radiation in which the tissue is placed, not the radiation absorbed.

The exposure to X- and gamma radiation is defined as the total charge of one sign produced by ionisation of a unit mass of air. Exposure is defined in terms of charge produced rather than energy deposited because it is far easier to measure small quantities of charge than it is to measure energy.

The unit of exposure is coulomb per kilogram ($C\,kg^{-1}$). A previous non-SI unit was the roentgen (R). This was named after one of the early pioneers of X-rays (Wilhelm Roentgen, or Röntgen, who won the first Nobel Prize for Physics in 1901 for his discovery of X-rays). 1 roentgen is equivalent to about $2.6 \times 10^{-4}\,C\,kg^{-1}$.

The concept of exposure can best be illustrated by the following example.

WORKED EXAMPLE 5.2

The energy required to produce one ion pair in air is 34 eV. Calculate the energy deposited in 1.0 g of air for an exposure of $0.03\,\mathrm{C\,kg^{-1}}$. The charge on the electron is $1.6 \times 10^{-19}\,\mathrm{C}$.

1.0 C of charge is equivalent to

$$\frac{1}{(1.6 \times 10^{-19})}\ \text{electrons} = 6.25 \times 10^{18}\ \text{electrons}$$

The energy required to produce these electrons in air is

$$\begin{aligned} 6.25 \times 10^{18} \times 34\,\mathrm{eV} &= 2.13 \times 10^{20}\,\mathrm{eV} \\ &= 2.13 \times 10^{20} \times 1.6 \times 10^{-19}\,\mathrm{J} \\ &= 34\,\mathrm{J} \end{aligned}$$

Therefore, for an exposure of $1.0\,\mathrm{C\,kg^{-1}}$, 34 J of energy is deposited in each kilogram of air.

Hence, the energy deposited in 1.0 g of air for an exposure of $0.03\,\mathrm{C\,kg^{-1}}$ is

$$34 \times 0.03 \times 1.0 \times 10^{-3} = 1.02 \times 10^{-3}\,\mathrm{J}$$

Absorbed dose

The **absorbed dose** D is measured as the mean energy absorbed per unit mass of tissue as a result of exposure to radiation. For mean energy E absorbed in mass M, then

$$D = \frac{E}{M}$$

The unit of absorbed dose is joule per kilogram ($\mathrm{J\,kg^{-1}}$). An absorbed dose of $1\,\mathrm{J\,kg^{-1}}$ is also referred to as an absorbed dose of 1 **gray** (Gy).

Absorbed dose is difficult to measure since the energy is absorbed in body tissue. However, as mentioned above, exposure can be measured relatively easily because it is the charge produced by ionisation per unit mass of air surrounding the tissue. Conversion charts are available so that any measured exposure can be translated into an absorbed dose in a particular medium or tissue, for a particular photon energy. The use of such conversions is illustrated by the graph of Figure 5.6 overleaf.

The variable $f/\mathrm{J\,C^{-1}}$ on the y-axis is the conversion factor and it relates dose $D/\mathrm{J\,kg^{-1}}$ to exposure $S/\mathrm{C\,kg^{-1}}$ by the expression:

$$D = f \times S$$

The conversion is complicated because, at low photon energies where the photoelectric effect predominates, the difference between the values of the proton numbers of air and of the tissue are important (attenuation coefficient depends on Z^3). Consequently, the value of f varies widely for different tissues. At higher photon energies where Compton scatter predominates, f assumes an almost constant value because attenuation is independent of Z value.

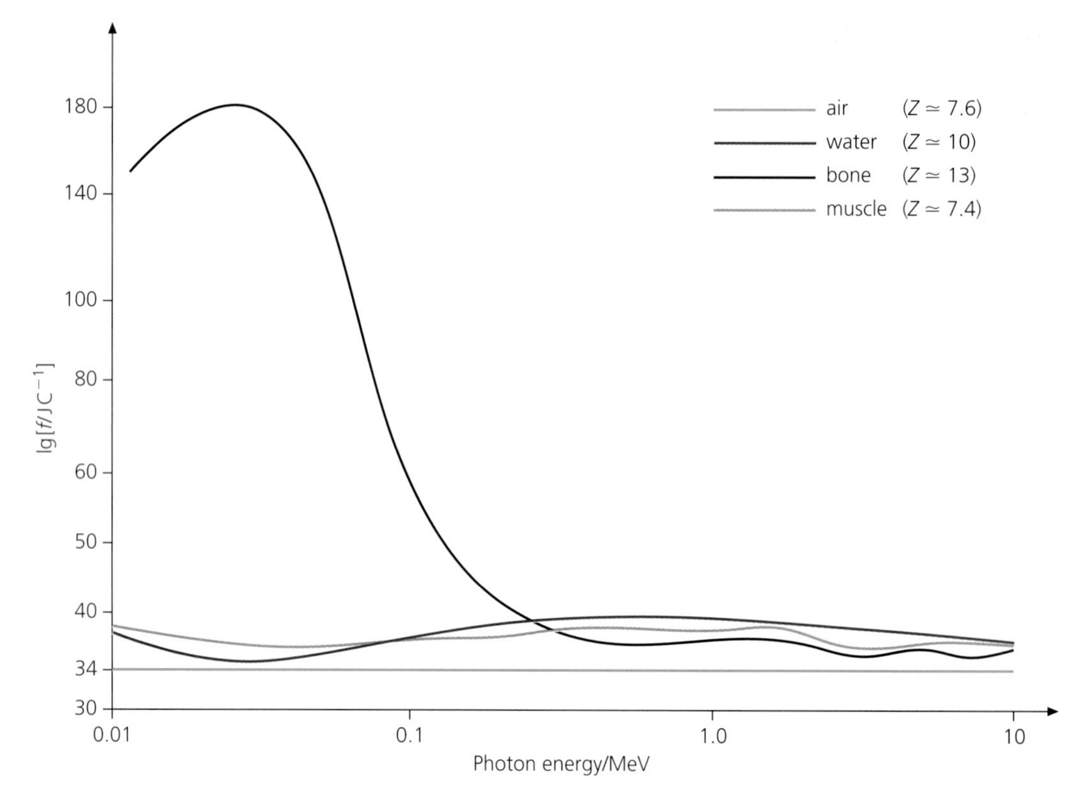

Figure 5.6 Use of a conversion factor *f* to calculate the absorbed dose from a measured exposure

Dose equivalent

The damage produced by the absorption of radiation depends not only on the amount of energy deposited, that is the dose, but also on its distribution. If the energy is deposited in a very small volume, then more damage is likely to be done than if it is well spread out. Alpha particles, with a comparatively large mass and charge $2e$, produce very dense ionisation and have a short path length. On the other hand, gamma radiation is much more penetrating and thus has a longer path length, producing much less ionisation per unit length of its path.

To allow for this effect, the dose D is multiplied by a **quality factor** Q, to obtain the **dose equivalent** H for a particular radiation:

$$H = Q \times D$$

Q is a number and does not have a unit. The basic unit of absorbed dose and dose equivalent are thus the same, i.e. $\mathrm{J\,kg^{-1}}$. In order to distinguish between the two, the unit of dose equivalent is also referred to as the **sievert** (Sv), where $1\,\mathrm{Sv} = 1\,\mathrm{J\,kg^{-1}}$. Some literature still refers to the rem. This is an old unit of dose equivalent, where $100\,\mathrm{rem} = 1\,\mathrm{Sv}$.

The quality factor cannot be given a precise value for each type of radiation but the generally accepted values are listed in Table 5.3.

Table 5.3 Quality factors

Type of radiation	Quality factor Q
Gamma rays, X-rays, beta radiation	1
Slow (thermal) neutrons	1–2
Fast neutrons, protons, alpha particles, heavy nuclei	10–20

Dose rate

The extent of the damage produced by radiation is dependent on the total dose received. It also depends on the rate at which the total dose is delivered. Biological cells are capable of some level of repair and some recovery is possible. This means that a particular dose delivered over a short period of time is likely to pose a greater risk of damage than the same dose spread out over a long period of time. When assessing the risk posed by radiation and setting limits to exposure, both total dose and dose rate must be considered.

Radiation levels and protection

In the early years of the development of the use of X-rays and the emissions from radioactive sources (at the beginning of the 20th century), the dangers associated with radiation were not appreciated. Even as late as 1960, X-ray machines were available in shoe shops so that a mother could see whether her child's shoes fitted properly! It is now beyond any doubt that radiation damages living tissue. Ideally the radiation dose should be zero, but this is impossible.

Background radiation

From the depths of space even to within the human body, there are sources of radiation giving rise to an absorbed dose.

- Cosmic radiation bombards the Earth and originates from nuclear reactions within stars.
- The Earth itself is radioactive since it contains radioactive elements. One substance of particular concern is the element radon. This gaseous radioactive element can collect in some badly ventilated buildings and, when breathed into the lungs, could cause cancer.
- Potassium-40 is a common radioactive element found in the body and is a main source of internal radiation.

The percentage of the natural background radiation dose received by the population of the UK as a result of different sources is shown in Figure 5.7 overleaf.

The figures are averages because living at higher altitude, or frequently travelling by air, means less protection from cosmic radiation. Similarly, living in areas where the underlying rocks are granite means increased concentrations of the radioactive elements uranium and radon.

Radiation monitoring

The detection and monitoring of radiation are of vital importance for radiological protection. Some detectors are considered in detail in other chapters: photographic film is widely used for the detection of X-rays, and emissions from radioactive sources are detected using Geiger counters and scintillation counters (see page 113). Besides the general monitoring of the working environment where radiation is present, it is important that each worker has a personal monitor that can be read on a regular basis. Numerous personal **dosimeters** (instruments for measuring radiation dose) are available, including pocket ionisation chambers and thermoluminescent dosimeters (see page 112). One that is in common use is the film badge (Figure 5.9) and this is described in detail in Chapter 8 (pages 110–112).

Figure 5.9 A radiographer wearing a film badge on his shoulder

The film badge is pinned to the clothing of the worker. Dependent on the potential for exposure to radiation, the film is processed and read after a period of 1–4 weeks. The degree of blackening of the film is measured to ascertain the radiation dose of each type of radiation. Film badges are cheap and easy to use and they require no maintenance. They also provide a permanent record of the dose. However, their accuracy is limited to 10–20% and the results are known only some time after the exposure. They do not give any indication of dose rate, only the total dose.

Questions

1 a State what is meant by a *photon* of electromagnetic radiation.

b Describe the mechanisms by which radiation causes damage to cells of living tissue.

c Suggest why alpha radiation presents little danger to health if the source of the radiation is outside the body but radon gas (an alpha-emitting source) is considered to be a serious health hazard.

2 The intensity I of a parallel beam of gamma radiation transmitted through a thickness x of an absorber is given by the expression

$$I = I_0 e^{-\mu x}$$

where I_0 is the intensity for $x = 0$.

a Identify the symbol μ.

b Explain what is meant by the *half-value thickness* (HVT) and show how this quantity relates to μ.

c Suggest the advantage of using the quantity μ/ρ, where ρ is the density of the absorber, rather than μ when discussing attenuation of X- or gamma-ray beams.

3 a Describe the mechanisms by which X- and gamma-ray photons lose energy in matter.

b State the dependence of each of the mechanisms in **a** on the proton number of the absorber.

c Sketch a graph showing the dependence on photon energy of the attenuation coefficient attributed to each mechanism in **a** and also their combined effect.

d Use your graph to explain why, for a photon energy of about 40 keV, there is a significant difference in the attenuation coefficients for bone and muscle.

4 a Distinguish between

i) exposure to radiation,

ii) absorbed dose,

iii) equivalent dose.

b The energy required to produce one ion pair in air is 34 eV. Calculate the energy deposited in 2.5 g of air for an exposure of 0.15 C kg^{-1}. (The charge on the electron is 1.6×10^{-19} C.)

5 a Explain what is meant by *background radiation* and outline the sources contributing to this radiation.

b Describe what is meant by *balanced risk* when referring to X-ray diagnosis.

6 a State the general philosophy of radiological protection agencies when considering maximum permissible doses of radiation.

b Suggest why maximum permissible doses of radiation vary according to

i) age and sex,

ii) occupation.

X-rays

In this chapter you will read about:

- the production of X-rays using a simple X-ray tube
- the X-ray tube spectrum
- the effect of a filter on the X-ray spectrum
- the nature of X-rays and their effect on matter
- the use of X-rays in imaging
- the use of X-rays in therapy

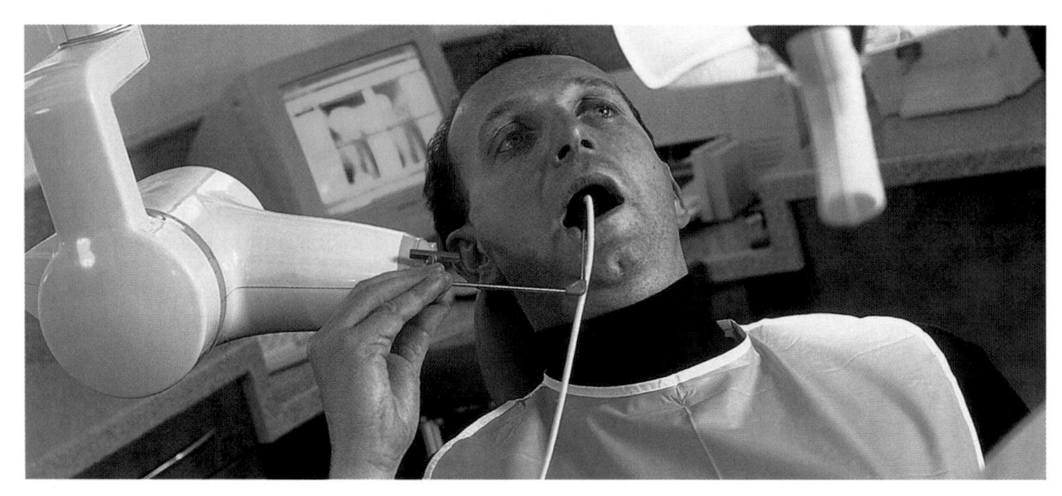

Figure 6.1 Use of an X-ray tube to obtain an image of the teeth inside the gums

X-rays in medicine

One of the principle uses for X-rays in medicine today is in the production of images of structures within the body. X-rays pass through a structure such as the leg and are then incident on undeveloped photographic film. X-rays penetrate soft tissue but are stopped by bone (see page 74). Photographic film is fogged when exposed to X-radiation and so when the film is developed a shadow photograph of the bone is obtained.

A second use for X-rays in medicine is in cancer therapy. X-rays are harmful to living tissue and can cause the death of cells. It is this property that is utilised when treating tumours.

The production of X-rays

X-ray photons are produced when fast-moving electrons undergo large decelerations by interaction with the atoms of a target metal. Figure 6.2 shows a simplified diagram of an X-ray tube.

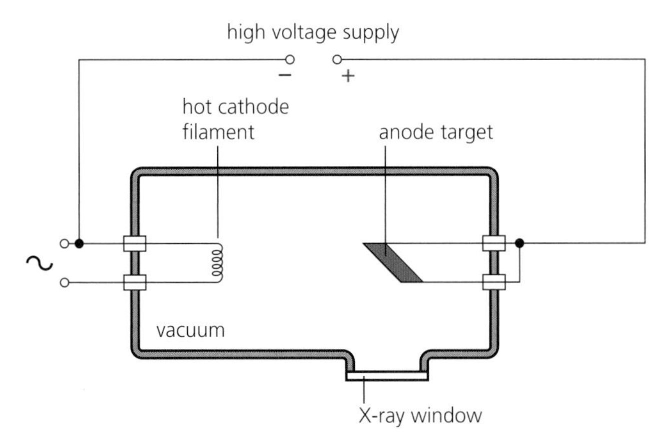

Figure 6.2 An X-ray tube

Electrons are produced at the heated cathode by **thermionic emission**. They are then accelerated through a potential difference, typically of the order of 30–100 kV, towards a target anode which is usually made of tungsten. The X-ray tube is evacuated, allowing the electrons to pass unhindered through the tube. The movement of electrons from cathode to anode constitutes a **tube current**. When the electrons are decelerated by hitting the target metal, less than 1% of the incident energy is converted into X-ray photons. The remainder of the energy is converted into thermal energy in the target metal. This produces a very large heating effect in the anode. To prevent the X-ray tube from overheating, oil is circulated around the tube. The oil removes heat from the tube by conduction and transfers it away through forced convection. Some X-ray tubes have a target anode that rotates. This exposes different parts of the target to the incident electron beam and hence distributes the thermal energy across the target metal.

WORKED EXAMPLE 6.1

An X-ray tube has a tube current of 40 mA due to a p.d. of 60 kV between the anode and cathode. X-ray photons are produced from the target metal with an efficiency of 0.65%. Calculate:

a the power consumption of the X-ray tube,

b the power of the X-ray beam.

a Electrical power $P = V \times I$

$$= 6.0 \times 10^4\,\text{V} \times 4.0 \times 10^{-2}\,\text{A}$$
$$= 2.4 \times 10^3\,\text{W} = 2.4\,\text{kW}$$

b As the efficiency of conversion of energy into X-ray photons is 0.65%, the power of the X-ray beam produced is:

$$\frac{0.65}{100} \times P = \frac{0.65}{100} \times 2.4 \times 10^3 = 15.6\,\text{W}$$

The energy of the X-ray photons produced

An electron of charge e coulombs accelerated through a potential difference of V volts gains an energy of eV. As the electrons in an X-ray tube hit the target they undergo a range of decelerations. The consequence is that X-ray photons of a range of energies are emitted from the tube. This radiation is called **bremsstrahlung** radiation. The maximum energy of an X-ray photon that emerges from the tube occurs when all of an electron's energy is converted to X-ray photon energy. Thus the maximum photon energy is eV. As the X-ray photons are electromagnetic in nature, the maximum photon energy may be equated to the electron energy eV as follows:

$$\text{maximum photon energy } E = eV = hf$$

where h is the Planck constant and f the frequency of the electromagnetic wave.

The minimum wavelength λ that corresponds to the maximum energy for the emitted X-rays may be calculated by substituting c/λ for f in the above equation:

$$eV = hf$$

$$eV = \frac{hc}{\lambda}$$

$$\lambda = \frac{hc}{eV}$$

The X-ray spectrum

Figure 6.3a shows the relative intensities of a typical distribution of X-ray photon energies from an X-ray tube of tube voltage 100 kV. Figure 6.3b shows the variation of relative intensity for the same distribution of photon energies but displayed as a function of the wavelength.

The lines labelled **K-lines** on the distribution graphs show high relative intensities at photon energies of 45–50 keV. They are formed due to the following sequence of events. After thermionic emission and acceleration across the X-ray tube, some electrons penetrate deep into the atoms of the target material and eject inner orbital electrons. Outer electrons drop into the vacancies formed with the subsequent emission of characteristic photons. These photons have a specific energy, the magnitude of which is equal to the difference between energy levels of the electron before and after it has filled the vacancy. The series of lines grouped at the lowest wavelength (and hence highest energy) are called the K-lines. These are due to transitions, or 'jumps', from energy levels above the ground state to the ground state (see Figure 6.4, page 88). Less energetic photons arising from jumps down to the first energy level from higher energy levels form another series of lines called the L-lines.

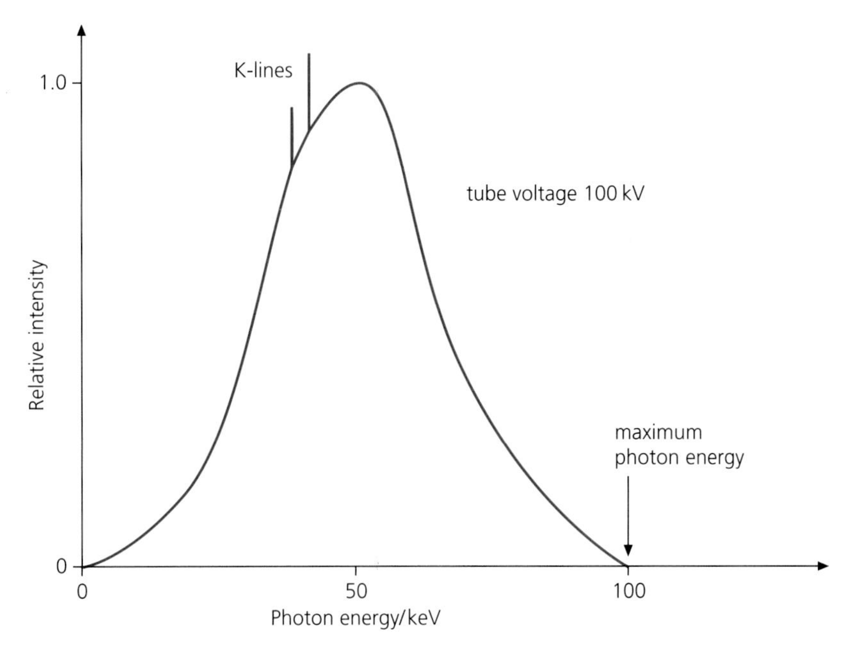

(a) As a function of photon energy

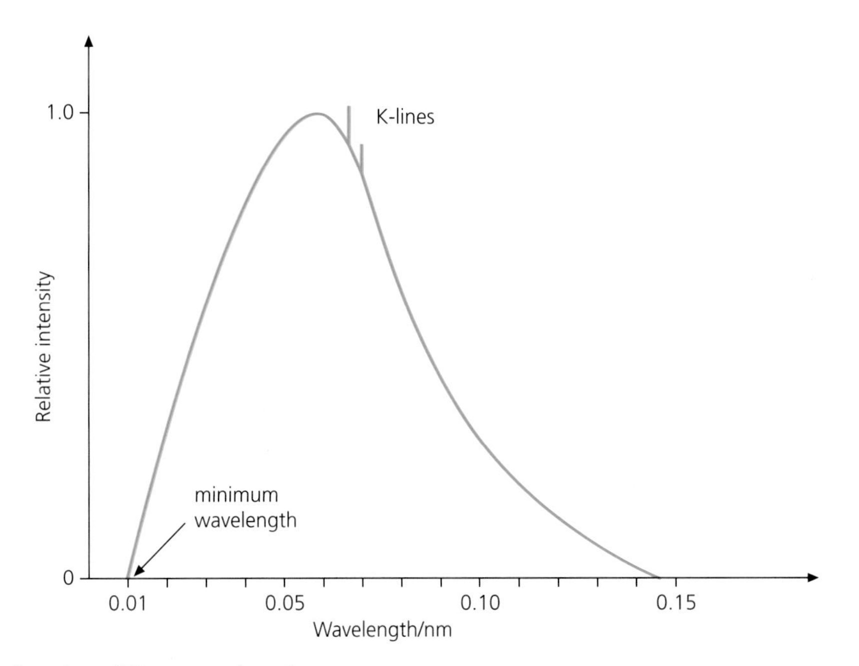

(b) As a function of X-ray wavelength

Figure 6.3 Typical X-ray spectra from an X-ray tube

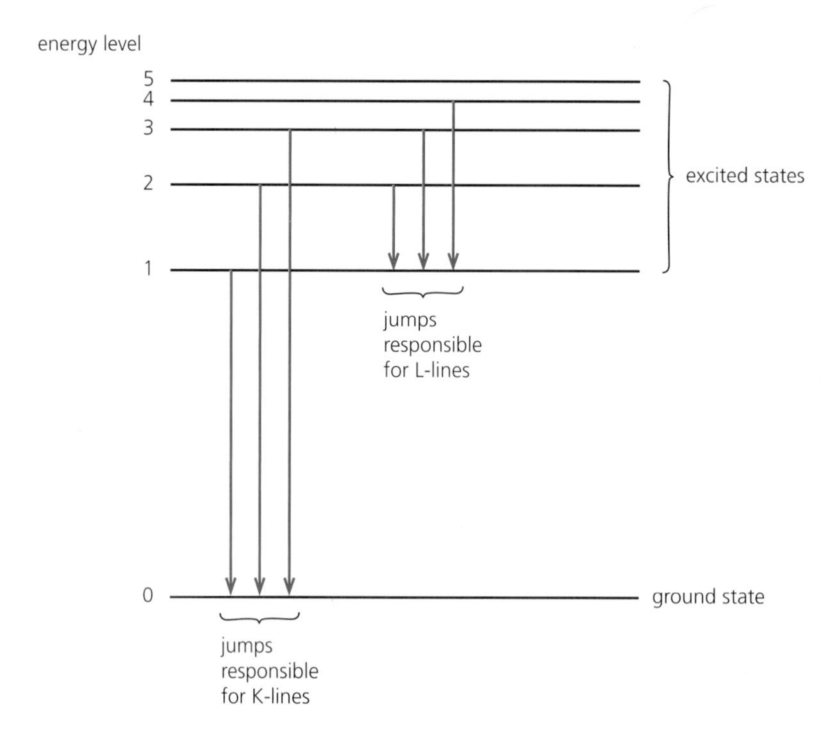

Figure 6.4 Electron transitions giving rise to lines in the X-ray spectrum

The use of X-rays in imaging

The principle of X-ray imaging relies upon individual parts of an absorbing medium **attenuating** the incident X-ray beam by different amounts. The amount of attenuation that occurs in a medium depends principally upon the thickness of the medium through which the X-rays pass, the atomic number of the atoms in the medium, and the energy of the X-ray photons incident on the medium. This is discussed in more detail in Chapter 5.

Filters

When X-rays are used to produce an image of an internal body structure, a patient is exposed to a range of X-ray photon energies. Low-energy X-ray photons are absorbed by the skin of the patient and play no part in the formation of the image. It is desirable, therefore, to remove these low-energy photons. This may be achieved by passing the X-ray beam through a filter such as a sheet of aluminium. The sheet absorbs a greater proportion of low-energy photons than high-energy photons and so increases the minimum energy of the photons to which the patient is exposed (see Figure 6.5).

Note that the peak of the spectrum has moved to a higher photon energy and the overall quantity of X-radiation has been reduced. This has the effect of making the X-ray beam more penetrating and the beam is said to have been **hardened**.

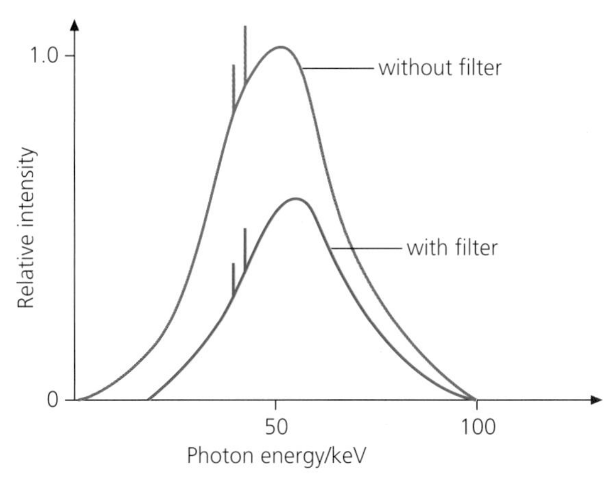

Figure 6.5 X-ray spectrum with and without a filter

WORKED EXAMPLE 6.2

In order to form an X-ray image of part of a patient, the lower-energy X-ray photons are required to be removed from an X-ray beam. The X-ray beam is thus passed through an aluminium filter of thickness 2.5 mm. The thickness of aluminium required to reduce the intensity of the X-ray beam by one half (the half-value thickness) is 1.5 mm. Calculate:

a the linear attenuation coefficient of aluminium for this beam,
b the intensity of the X-ray beam on emerging from the 2.5 mm filter if the incident intensity is $30\,\text{MW}\,\text{m}^{-2}$.

a $\dfrac{I}{I_0} = e^{-\mu x}$ where μ is the linear attenuation coefficient (see page 72).

Putting in the half-value thickness:

$$\tfrac{1}{2} = e^{-\mu \times 1.5}$$

$$\ln 0.5 = -\mu \times 1.5$$

$$\mu = -\frac{\ln 0.5}{1.5}$$

$$= 0.46\,\text{mm}^{-1}$$

b $I_0 = 30 \times 10^6\,\text{W}\,\text{m}^{-2}$

$$I = I_0 \times e^{-\mu x}$$
$$= 30 \times 10^6 \times e^{-0.46 \times 2.5}$$
$$= 9.5 \times 10^6\,\text{W}\,\text{m}^{-2}$$

The effect on an X-ray image of a variation in the proton number of the absorbing medium

The earlier explanation of X-ray images at the beginning of the chapter (page 84) was a simplification. While essentially an X-ray photograph is a 'shadow photograph', there are shades of grey between the black and white, formed where the X-ray beam has been partially attenuated in the medium. There is a greater attenuation of X-rays in bone than in muscle and a greater attenuation of X-rays in muscle than in fat. The average proton number (atomic number) for bone is larger than that for muscle tissue, which in turn is larger than that for fat.

In order to consider how a variation in the proton number affects the attenuation of an X-ray beam in a medium, the effect of this variation on the linear attenuation coefficient will be discussed. When the value of the linear attenuation coefficient μ for a given medium and a given X-ray beam is large, the attenuation within the absorbing medium is also large (see page 72).

For the attenuation mechanism *simple scatter*, the linear attenuation coefficient varies approximately as Z^2 (the square of the proton number; see Table 5.1 on page 74). In other words, as the proton number of the absorbing medium doubles, the linear attenuation coefficient increases by a factor of four. This mechanism enables differentiation between tissues like muscle, with low proton numbers, and tissue like bone of higher proton number. The result on a photographic film that is exposed to X-radiation emerging from a patient is a contrast in intensity between X-rays that have passed through different tissues.

Where contrast is required in tissues that have similar proton numbers, a **contrast medium** such as barium or iodine is employed. A contrast medium is a substance that has a high proton number. When injected into the blood, the contrast medium absorbs X-rays to a greater degree than the surrounding tissue. The throat, stomach or intestines may be imaged using the same principle with the use of the contrast medium barium, taken orally (swallowed); see Figure 6.6.

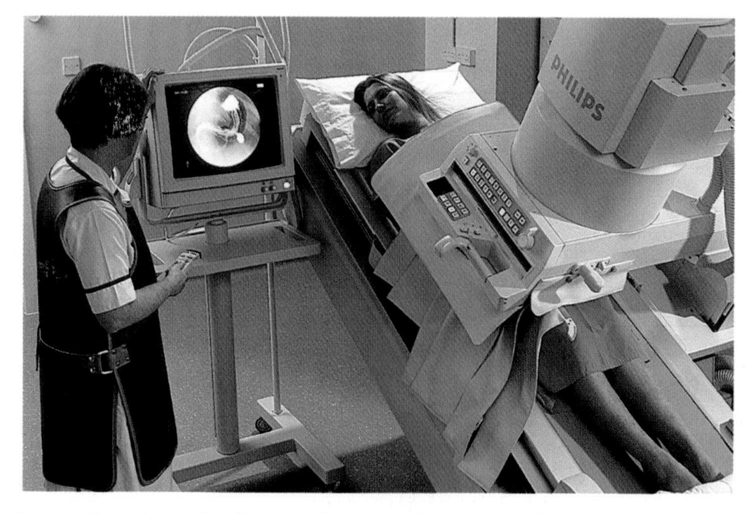

Figure 6.6 Patient undergoing a 'barium meal' X-ray of the stomach

For the *photoelectric effect*, the linear attenuation coefficient varies approximately as Z^3. This mechanism differentiates between soft tissue and bone to a much greater degree than simple scatter, and is the principal mechanism in ordinary X-ray imaging.

For the mechanism *Compton scatter*, the linear attenuation coefficient is independent of Z. This mechanism does not contribute at all to the differentiation of tissue type by proton number and merely adds to the radiation dose absorbed by the patient.

For *pair production*, the linear attenuation coefficient varies approximately as Z^2. As the energies required for pair production are so large this mechanism is not employed in X-ray imaging.

The effect of the energy of the X-ray photons

Typical X-ray tube voltages for imaging purposes are in the range 80–100 kV. This gives predominant photon energies of about 30–50 keV. The photoelectric effect is the dominant attenuation mechanism at these photon energies (see Table 5.1) and so a good contrast between bone and soft tissue is gained. Simple scatter, which reduces the quality of an image, makes only a small contribution to the overall attenuation at these energies, while pair production does not occur and so makes no contribution at all. Compton scatter occurs but makes no contribution to the differentiation of tissue types, as attenuation by this method is independent of any variation in Z.

When X-rays are used for therapy rather than imaging, the photon energies employed are usually in the range 0.5–5 MeV. This is because, at the lower photon energies used for imaging, bone will absorb up to eleven times more incident energy than will the surrounding tissue, so more damage is done to the bone than to the tumour being treated. For photon energies above 0.5 MeV Compton scatter is the dominant attenuation mechanism. This is independent of the proton number of the absorbing medium and so there is no preferential absorption by bone.

The use of X-rays in therapy

When X-rays interact with matter, ionisation results (see Chapter 5). This may affect the function of molecules such as DNA and may cause cell death. Cells that are in the process of dividing are most susceptible to damage by X-rays. As cancerous cells divide at a greater rate than healthy cells, exposure to a dose of X-rays will kill cancerous cells at a greater rate than healthy cells. This is the basis for cancer therapy.

Reducing damage to healthy cells

Whenever X-rays are used for therapy, healthy cells are killed. In deciding the course of treatment for a patient, it is important to consider minimising the damage to the healthy tissue that surrounds the cancer. There are a number of ways in which this may be achieved. The tumour is accurately located and then X-rays are aimed at the tumour from different directions (Figure 6.7).

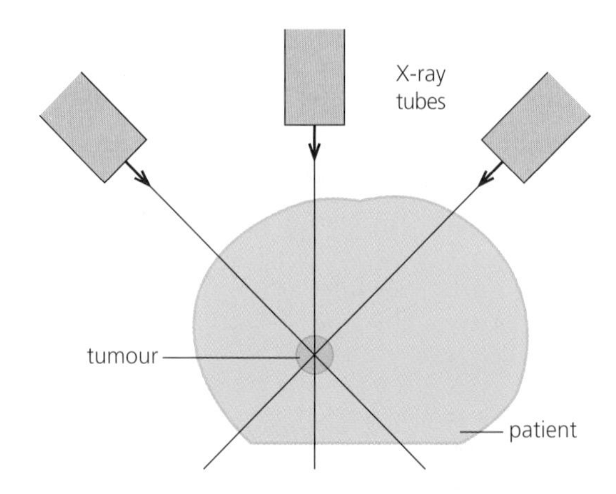

Figure 6.7 Multiple beam therapy

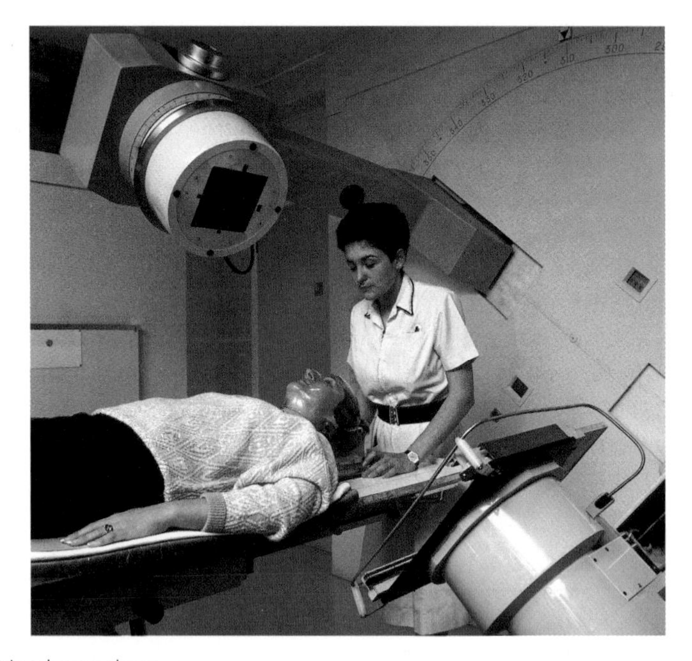

Figure 6.8 Rotating beam therapy

The dose at the tumour from all of the X-ray beams is cumulative and much greater than the dose received by the surrounding tissue. This method of treatment is called **multiple beam therapy** and is usually used for deep-seated tumours.

A variation on this method of treatment involves the X-ray tube rotating about the patient with the tumour at the centre of rotation (Figure 6.8). Once again the tumour receives a much larger dose compared to the surrounding tissue.

X-ray therapy is usually given in small doses called **fractions** over a period of days or over several weeks. Healthy cells recover at a greater rate than cancerous cells and so the period between treatment gives the tissue surrounding the tumour time to recover.

Questions

1 **a** Describe the production of an X-ray beam, outlining the various energy conversions that take place in a simple X-ray tube.

b Explain why a distribution of X-ray photon energies is obtained from a typical X-ray tube.

c An X-ray tube has a tube voltage of 90 kV. Calculate the maximum energy of the X-ray photons (in J) that emerge from the tube. ($e = 1.6 \times 10^{-19}\,C$)

d Calculate the minimum wavelength of the X-ray photons from this X-ray tube. ($h = 6.6 \times 10^{-34}\,J\,s$, $c = 3.0 \times 10^{8}\,m\,s^{-1}$)

2 **a** Sketch a graph of relative intensity (*y*-axis) against X-ray photon energy (in keV) for an X-ray tube of tube voltage 80 kV.

b The X-ray beam is passed through an aluminium filter. Sketch, on the same axes as in **a**, a graph to show how the relative intensity of the X-ray beam that emerges from the aluminium varies with photon energy.

c Explain the purpose of using such a filter.

d The half-value thickness of aluminium for this X-ray tube at the tube voltage of 80 kV is 2.2 mm. Calculate the linear attenuation coefficient for this X-ray beam.

3 **a** Explain the following attenuation mechanisms:

i) *the photoelectric effect*,

ii) *simple scatter*.

b Explain why the mechanisms in **a** are useful in the production of an X-ray image of a broken bone, whereas *Compton scatter* is not useful for this purpose.

c The photoelectric effect may not be a suitable mechanism in the treatment of a tumour. Give two reasons why Compton scatter might be a preferable mechanism.

4 The p.d. across an X-ray tube is 70 kV. The X-ray tube current is 50 mA. The efficiency of the X-ray tube is 0.95%. The diameter of the emergent X-ray beam is 1.0 mm.

a Calculate the power consumption of the X-ray tube.

b Calculate the power of the X-ray beam.

c Calculate the intensity of the emerging X-ray beam.

d Calculate the thickness of a copper filter that will cause the intensity of this X-ray beam to fall to 0.125 of the initial intensity. (The linear attenuation coefficient for this X-ray beam in copper is $693\,m^{-1}$.)

5 **a** Explain why X-rays may be employed to treat a tumour.

b Explain two methods used during therapy to minimise damage to the healthy tissues that surround a tumour.

Radioisotopes

In this chapter you will read about:

- some of the terms associated with radioactivity
- examples of how radionuclides are produced
- the use of radioisotopes in diagnosis
- the use of radioisotopes in therapy

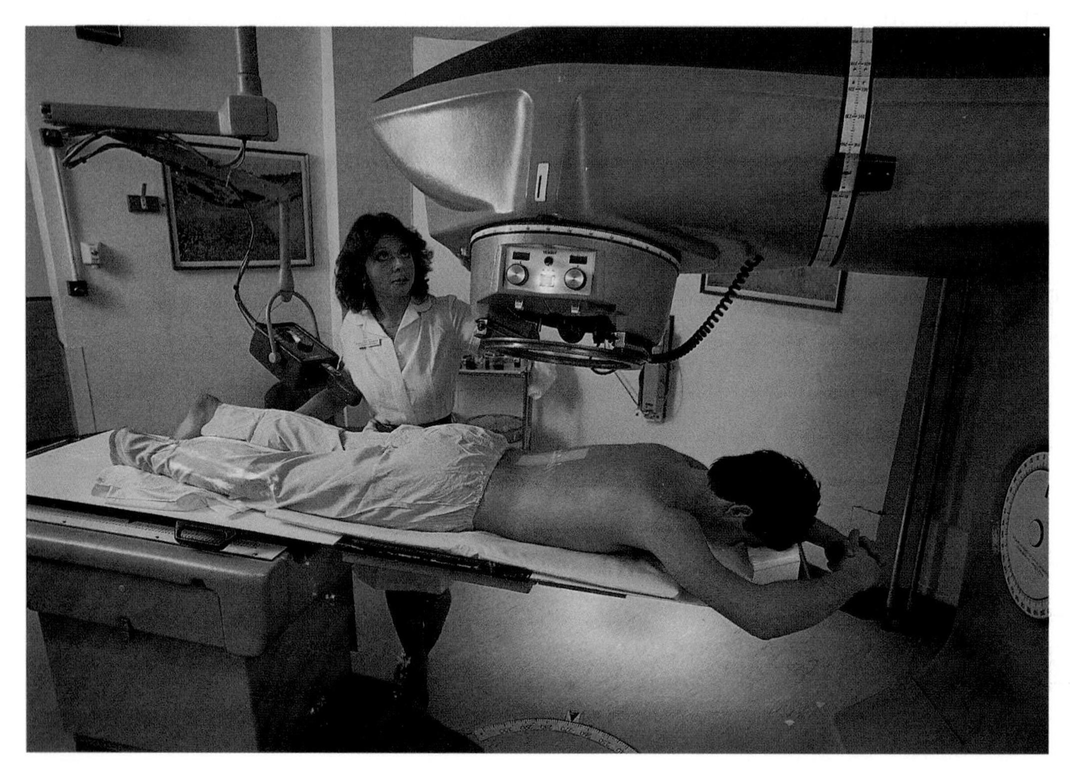

Figure 7.1 Patient undergoing radiotherapy

Introduction

Any atom of an element can be represented symbolically as

$$^{A}_{Z}X$$

where X is the chemical symbol for the element
Z is the proton number (atomic number)
and A is the nucleon number (mass number).

Thus, an atom of the element X has Z protons in its nucleus, together with $(A - Z)$ neutrons. Furthermore, if the atom is neutral, there will be Z electrons in orbitals around the nucleus.

Each type of nucleus is called a **nuclide** and this means that any nuclide has specific values for Z and A. Nuclides with the same proton number but with different nucleon numbers are referred to as **isotopes**, for example $^{12}_{6}C$ (carbon-12) and $^{14}_{6}C$ (carbon-14). Some isotopes are given specific names – for example, deuterium is hydrogen-2 ($^{2}_{1}H$) and tritium is hydrogen-3 ($^{3}_{1}H$).

When the ratio of protons to neutrons in a nucleus goes outside a certain range, the nucleus becomes unstable. It decays to become more stable and, in so doing, emits radiation. Such isotopes are said to be **radioactive** and are referred to as **radioisotopes**. They emit either alpha particles (α-radiation) or beta particles (β-radiation) to become isotopes of different elements. Usually, gamma radiation (γ-radiation) is emitted together with the α- or β-radiation, as high-energy photons. The gamma emission reduces the energy of the nucleus without changing its nuclear composition.

When a nucleus decays spontaneously, that is without any outside influence, into a more stable nucleus, the process is called natural radioactivity. Stable nuclei can be induced to become radioactive by bombarding them with nuclear particles in nuclear reactors or particle accelerators (see pages 101–102).

Radioactive decay

It has already been pointed out that radioactive decay is spontaneous. This means that the rate at which nuclei decay cannot be changed by altering the environmental conditions. The decay is also random, depending only on the particular nuclide. Randomness means that it can never be predicted which particular nucleus will decay next or at what time a nucleus will decay. However, the nuclei of any particular isotope each have a constant probability (chance) of decay per unit time. This constant probability of decay per unit time is known as the **decay constant** λ of the nuclide.

If a sample of radioactive nuclei of a particular isotope contains N nuclei at time t and $(N - dN)$ at time $(t + dt)$ (the nuclei are decaying, so N decreases as time increases and thus dN must be subtracted from N), then

$$\text{probability of decay} = -\frac{dN}{N}$$

and

$$\text{probability of decay per unit time} = -\frac{dN/N}{dt}$$

This quantity is equal to the decay constant λ. Re-arranging,

$$\frac{dN}{dt} = -\lambda N$$

The quantity $-(dN/dt)$ is the **activity** A of the sample. That is, it is the value of the rate of decay of nuclei at that time. Activity has the unit **becquerel** (Bq) where 1 becquerel (Bq) = 1 disintegration per second (s^{-1}).

For a sample containing N_0 nuclei and having an activity A_0 at time $t = 0$, and N nuclei and activity A at time t, the equation $dN/dt = -\lambda N$ can be solved to give the equations:

$$N = N_0 e^{-\lambda t}$$

and

$$A = A_0 e^{-\lambda t}$$

These equations for the number of nuclei and for the activity may be represented graphically as in Figure 7.2. The change illustrated by the graph is an exponential change and in this case it is an **exponential decay**. The number of nuclei and the activity both approach zero as time increases. However, since it is not possible to tell when the number or activity reaches zero, the 'life' of the radioactive nuclide cannot be determined. Instead, a fundamental property of an exponential decay curve is used to define what is called the **half-life** of the isotope.

It can be seen in Figure 7.2 that it always takes the same length of time for the number of nuclei or the activity to halve, regardless of the starting point. Thus the half-life of a radioisotope is defined as the average time taken for the activity, or the number of nuclei of that particular isotope, to be reduced to one half of its initial value.

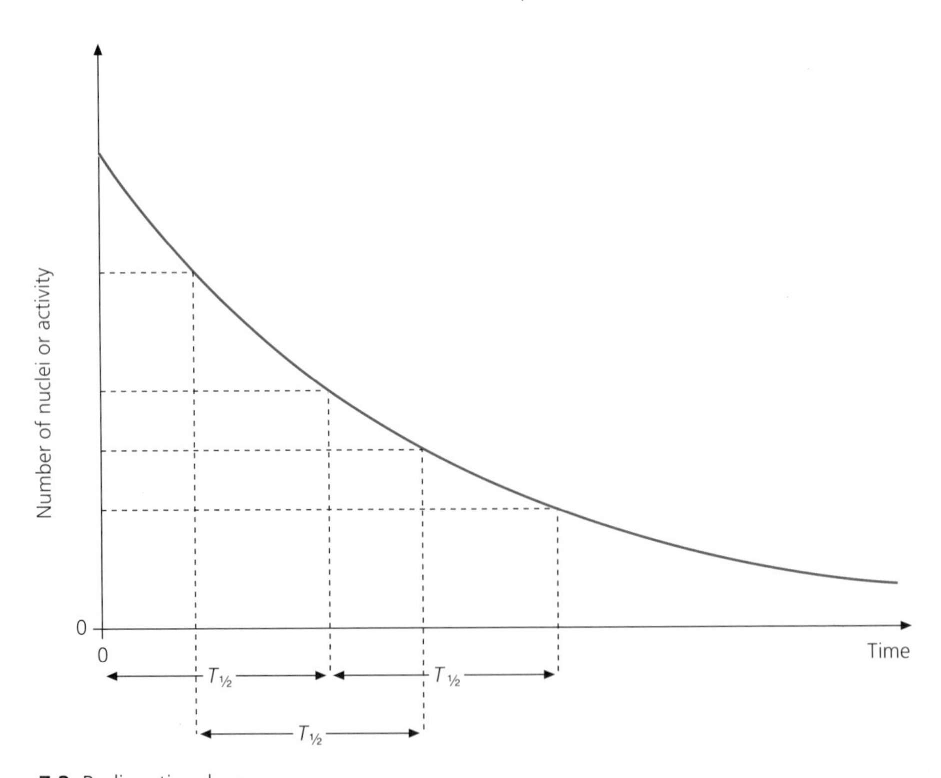

Figure 7.2 Radioactive decay

Half-life is given the symbol $T_{\frac{1}{2}}$. Referring to the equation $N = N_0 e^{-\lambda t}$, when $N = \frac{1}{2}N_0$ then $t = T_{\frac{1}{2}}$ and we have:

$$\frac{1}{2}N_0 = N_0 e^{-\lambda T_{\frac{1}{2}}}$$
$$\frac{1}{2} = e^{-\lambda T_{\frac{1}{2}}}$$
$$2 = e^{\lambda T_{\frac{1}{2}}}$$

Taking logarithms to base e, we obtain:

$$\ln 2 = \lambda T_{\frac{1}{2}}$$

This is an important expression because it links half-life, a quantity that can be determined experimentally, with the decay constant.

Values of the decay constant and the half-life for some isotopes used in medicine are shown in Table 7.1.

Table 7.1 Some commonly used radioisotopes

Radioisotope		**Decay constant λ**	**Half-life $T_{\frac{1}{2}}$**
Tritium	$^{3}_{1}H$	0.0564 year^{-1}	12.3 year
Carbon-14	$^{14}_{6}C$	1.24×10^{-4} year^{-1}	5570 year
Phosphorus-32	$^{32}_{15}P$	0.0475 day^{-1}	14.6 day
Potassium-40	$^{40}_{19}K$	4.65×10^{-10} year^{-1}	1.49×10^{9} year
Iodine-125	$^{125}_{53}I$	0.0116 day^{-1}	60.0 day
Iodine-131	$^{131}_{53}I$	0.0861 day^{-1}	8.05 day
Radon-220	$^{220}_{86}Rn$	0.0127 s^{-1}	54.5 s
Radium-226	$^{226}_{88}Ra$	4.28×10^{-4} year^{-1}	1620 year

Biological half-life

When a chemical is taken into the body, its concentration decreases with time. The chemical may be broken down within the body but eventually it is removed from the body by normal processes such as respiration, urination and defecation. The concentration of the chemical (which may be a prescribed drug) decreases exponentially with time and therefore it has a half-life, known as the **biological half-life** T_B. This means that, if a radioisotope with a **physical half-life** T_P is introduced into the body, its concentration will decrease exponentially due to both biological and physical processes. This results in an **effective half-life** T_E, which determines the reduction in activity of the radioisotope in the body, given by:

$$\frac{1}{T_E} = \frac{1}{T_B} + \frac{1}{T_P}$$

It should be remembered that the physical half-life is a constant for any particular radioisotope. However, the biological half-life depends on the state of health of the individual and the organ in which the radioisotope is located.

WORKED EXAMPLE 7.1

Sodium-24 has a half-life of 15 hours. A freshly prepared sample of sodium-24 has an initial activity of 4.6×10^7 Bq. Given that the Avogadro constant is $6.0 \times 10^{23}\,\text{mol}^{-1}$, determine:

a the decay constant, in s^{-1}, of sodium-24,
b the initial mass of the sample,
c the activity of the sample after 2.0 days.

a $\ln 2 = \lambda T_{\frac{1}{2}}$

so $$\lambda = \frac{\ln 2}{T_{\frac{1}{2}}} = \frac{0.693}{15 \times 60 \times 60\,\text{s}}$$
$$= 1.3 \times 10^{-5}\,\text{s}^{-1}$$

b $\text{Activity} = -\left(\frac{dN}{dt}\right) = \lambda N$

so $$N = \frac{\text{activity}}{\lambda} = \frac{4.6 \times 10^7\,\text{s}^{-1}}{1.3 \times 10^{-5}\,\text{s}^{-1}}$$
$$= 3.5 \times 10^{12}$$

24 g of sodium-24 contains 6.0×10^{23} nuclei

so $$\frac{\text{mass of sample}}{N} = \frac{24\,\text{g}}{6.0 \times 10^{23}}$$

and $$\text{mass of sample} = \frac{24\,\text{g} \times 3.5 \times 10^{12}}{6.0 \times 10^{23}}$$
$$= 1.4 \times 10^{-10}\,\text{g}$$

c $A = A_0 e^{-\lambda t}$

After $t = 2$ days $= 2 \times 24 \times 3600$ s:

$$A = 4.6 \times 10^7 \exp(-1.3 \times 10^{-5} \times 2 \times 24 \times 3600)\,\text{Bq}$$
$$= 4.9 \times 10^6\,\text{Bq}$$

Note the use of the term 'exp'. This is used in some textbooks as an alternative to writing a number to the power of e. Thus, e^{kx} would be written as $\exp(kx)$.

WORKED EXAMPLE 7.2

Iodine-131 has a physical half-life of 8.0 days. It is removed from the body with a biological half-life of 21 days. Calculate:

a the effective half-life of the isotope,

b the time taken for the activity within the body to fall to 10% of its initial value.

a

$$\frac{1}{T_E} = \frac{1}{T_B} + \frac{1}{T_P}$$

$$= \frac{1}{21} + \frac{1}{8}$$

giving

$$T_E = 5.8 \text{ days}$$

b Activity $A = A_0 e^{-\lambda t}$

and

$$\lambda = \frac{\ln 2}{T_{\frac{1}{2}}}$$

$$= \frac{\ln 2}{5.8} = 0.12 \text{ day}^{-1}$$

We have

$$\frac{A}{A_0} = 0.10 = e^{-0.12t}$$

giving

$$t = 19 \text{ days}$$

The choice of radioisotopes for use in medicine

Radioisotopes may be used for two distinct purposes.

1. **Diagnosis.** The functioning of organs within the body may be assessed and tumours may be located. The radiation from an external gamma-ray source is directed at the body. Radiation passing through the body causes a flash of light in a scintillator crystal and is detected by means of photomultiplier tubes. The tumour would result in a different intensity of radiation penetrating the body. Alternatively, the source may be a radioisotope that has been introduced into the body, and its activity in the affected area is monitored.
2. **Therapy.** Radiation, again either from an external source or from one inside the body, is used to destroy cancerous cells.

Whether for diagnostic or therapeutic purposes, if a radioisotope is to be taken into the body (ingested) then consideration must be made of the following:

- chemical suitability
- biological behaviour
- physical properties.

Chemical suitability

As for any other pharmaceutical product, the chemicals must have the required level of purity. The additional problem here is that the product is radioactive and may not have a long half-life. Not only is preparation time a consideration but also any storage time will reduce the activity.

Preparation of a radio-pharmaceutical product is in two stages. First, the radioisotope has to be produced (this will be discussed on pages 101–104), separated chemically from any other material present and then purified. Second, the radioisotope must be bonded into molecules as a pharmaceutical product. The pharmaceutical product is then said to be 'labelled', in that the presence of the substance can be detected by the emission of radiation from the bonded radioisotope.

Biological behaviour

Generally, when a medicine is taken, the intention is to change some function in the body. For example, we take a painkiller if we have toothache. This affects the function of the nerves from the tooth. However, if a radio-pharmaceutical is to be used for diagnosis, then it is important that the drug does not affect the functioning of the organ or the metabolic pathway that is being assessed. The chemical should simply accumulate in the organ to be studied.

As an example, one of the longest established applications of nuclear medicine is the use of iodine-131 in the thyroid gland. The thyroid gland requires iodine in order to produce a hormone called thyroxine. Iodine-131 introduced into the body (in a solution of sodium iodide) is absorbed into the thyroid gland, rather than surrounding tissues. Monitoring radiation levels in the region of the neck then allows the size of the gland to be assessed. The uptake and subsequent excretion of the iodine in the gland over the next two days allows for its rate of functioning to be estimated. If very large doses of iodine are administered, the radiation may be used to kill cancerous cells in the thyroid.

Physical properties

For diagnosis, the radioisotope must emit radiation of the correct type and energy so that it can be detected externally. In the case of therapy, the type of radiation and its energy must be chosen so that the radiation has its most harmful effects in the cancerous cells. The intention is to cause as little radiation damage to healthy cells as is possible. The radiation dose both for diagnosis and for therapy should be minimised.

The type and energy of the radiation are not the only factors. Once the radio-pharmaceutical has carried out its function, then it is undesirable for it to remain in the body. The effective half-life of the pharmaceutical has to be chosen carefully. If it is too short, the count rate will be too low before the diagnosis or therapy is complete. If it is too long, there is unnecessary absorbed dose.

Some examples of radioisotopes used in diagnosis and therapy are shown in Table 7.2.

Table 7.2 Some radioisotopes and their uses in medicine

Radioisotope		Uses
Tritium Sodium-24 Potassium-42 Iodine-131	^{3}H ^{24}Na ^{42}K ^{131}I	Body composition: volumes of body fluids and concentrations of ions may be assessed
Iodine-123 Xenon-133	^{123}I ^{133}Xe	Heart and lungs: cardiac output, blood circulation studies and respiration studies
Phosphorus-32 Technetium-99m Iodine-123	^{32}P $^{99}Tc^{m}$ ^{123}I	Diagnosis: detection and location of tumours
Cobalt-60 Iodine-131 Phosphorus-32	^{60}Co ^{131}I ^{32}P	Therapy: used to destroy cancerous cells, as either an external or internal source of ionising radiation

The production of radioisotopes

Although the naturally occurring isotope radium-226 still has some use in medicine, the vast majority of radioisotopes now used are produced artificially. These isotopes are produced by bombarding stable isotopes either with neutrons in a nuclear reactor or with high-energy charged particles in an accelerator.

Nuclear reactor methods

In a nuclear reactor the fission of each uranium-235 nucleus produces, on average, two or three neutrons. If materials such as phosphorus-31, sodium-23 or cobalt-59 are placed in the reactor, they will absorb neutrons produced in the fission reactions and become radioactive isotopes. For example, phosphorus-31 absorbs a neutron to become phosphorus-32. The nuclear equation is:

$$^{31}_{15}P + ^{1}_{0}n \rightarrow ^{32}_{15}P + \gamma$$

The phosphorus-32 then undergoes β^- decay with a half-life of about 15 days. Since it is not possible to separate the phosphorus-32 from the phosphorus-31 and also because only a small fraction of the phosphorus-31 will capture neutrons, the specific activity (the activity per unit mass) of the sample after bombardment will not be high.

The capture of a neutron may, if the energy of the incident neutron is high, lead to the ejection of a proton from the target nucleus. An example of this is the production of carbon-14 from nitrogen-14:

$$^{14}_{7}N + ^{1}_{0}n \rightarrow ^{14}_{6}C + ^{1}_{1}p$$

Carbon-14 is used for labelling molecules in medical research.

Tritium (an isotope of hydrogen) is also used in medicine. This is produced by the bombardment of lithium-6 with high-energy neutrons, resulting in the production of a tritium nucleus and an alpha particle:

$$^{6}_{3}Li + ^{1}_{0}n \rightarrow \underset{\text{tritium}}{^{3}_{1}H} + \underset{\text{alpha}}{^{4}_{2}He}$$

Accelerator methods

Although the majority of radioisotopes are produced in nuclear reactors, accelerators are used for producing isotopes that have short half-lives. The technique is more expensive than using nuclear reactors, but some large hospitals do have the facilities.

The method involves bombarding nuclei with charged particles such as deuterons, alpha particles or protons. Since the particles are positively charged, the bombarding particle must have a high energy (of the order of MeV) so that it can penetrate the nucleus. These energies are achieved using a particle accelerator such as a **cyclotron** (Figure 7.3).

As an example, iodine-123 is produced by bombarding antimony-121 with alpha particles:

$$^{121}_{51}\text{Sb} + ^{4}_{2}\text{He} \rightarrow ^{123}_{53}\text{I} + 2^{1}_{0}\text{n}$$

Iodine-123 is sometimes preferred to iodine-131 because it emits only gamma photons and has a half-life of 13 hours. Iodine-131 emits beta particles as well as gamma photons, and has a half-life of 8 days.

Figure 7.3 Cyclotron used for producing medical radioisotopes

The technetium generator

This generator enables technetium-99m ($^{99}Tc^{m}$), having a half-life of 6 hours, to be produced without the need for an on-site nuclear reactor or cyclotron. The 'm' denotes that $^{99}Tc^{m}$ is in a metastable state. That is, each nucleus will emit a low-energy gamma photon to become technetium-99. Technetium-99 itself is a β^- emitter with a half-life of 2.2×10^5 years and is thus unsuitable for medical use.

Molybdenum-99 is produced in an off-site nuclear reactor. This is absorbed on alumina in a glass or plastic column in the form of ammonium molybdenate. The molybdenum decays by β^- emissions with a half-life of 67 hours and forms technetium-99m as pertechnate ions in the column. The column (Figure 7.4) is then a ready source of technetium-99m.

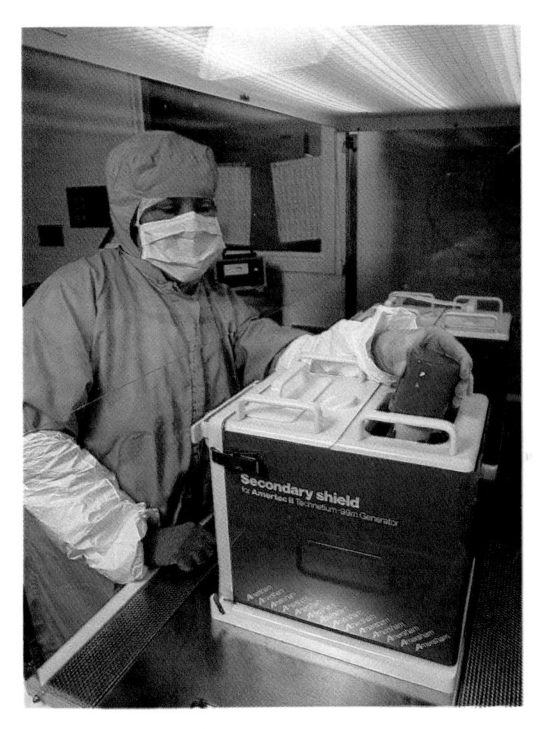

Figure 7.4 Removing a technetium column from a generator

When sodium chloride solution is passed through the column, the molybdenate ions are not affected by the chloride ions but the pertechnate ions exchange places with the chloride ions. This produces sodium pertechnate in the sodium chloride (saline) solution. This solution may be made compatible with blood.

After removal of the pertechnate ions in the column, the concentration of the $^{99}Tc^{m}$ begins to build up again as the molybdenum disintegrates. When the concentration of $^{99}Tc^{m}$ has built up again, the column can be re-used. The activity of the $^{99}Tc^{m}$ within the column is illustrated in Figure 7.5.

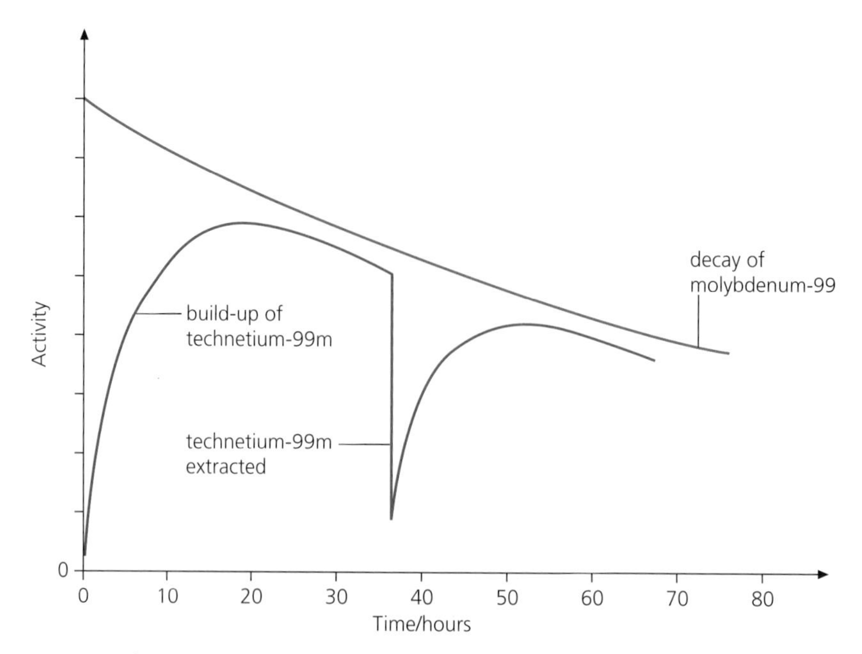

Figure 7.5 Activity of technetium-99m in the generator column

Note that, due to the decay of the parent isotope molybdenum-99, there is a finite (and practical) limit to the number of samples of technetium that can be extracted.

WORKED EXAMPLE 7.3

A sample of molybdenum in a technetium-99m generator has a half-life of 67 hours and an initial activity of 5.0×10^9 Bq. Technetium-99m is extracted from the generator at 48-hour intervals. At each extraction, 85% of the activity is removed.

Calculate the number of extractions that can be made if the activity of the extracted sample is not to fall below 5% of the initial activity of the molybdenum.

Minimum activity of technetium $= 0.05 \times 5.0 \times 10^9 = 2.5 \times 10^8$ Bq

This activity in the extracted technetium is 85% of the activity of the molybdenum at the time of extraction. So:

$$\text{minimum activity of molybdenum} = \frac{100}{85} \times 2.5 \times 10^8\ \text{Bq}$$

$$= 2.94 \times 10^8\ \text{Bq}$$

Activity $A = A_0 e^{-\lambda t}$ and $\lambda = \dfrac{\ln 2}{T_{\frac{1}{2}}}$.

We have $\lambda = 0.0103$ hour^{-1} and so the time t to reach the minimum activity is given by:

$$2.94 \times 10^8 = 5.0 \times 10^9 \times e^{-0.0103t}$$

giving

$$t = 275 \text{ hours}$$

Extractions are taken at 48-hour intervals. 275/48 = 5.7, therefore 5 extractions can be made.

Examples of the use of radioisotopes

1 Diagnosis

The thyroid

The use of iodine-131 for the assessment of the functioning of the thyroid gland (see page 100) is a classic example of where the normal metabolic processes of the gland concentrate the radioisotope within itself. Iodine-123 is now preferred to iodine-131, and technetium-99m is frequently used as an alternative. Technetium-99m is taken up by the thyroid in the same way as iodine, but it is released more easily. This means that the overall radiation dose received in the thyroid can be reduced.

The location of the radioisotope and its concentration are determined using scanners (see Chapter 8). The scan image enables the size of the gland to be determined and also its activity.

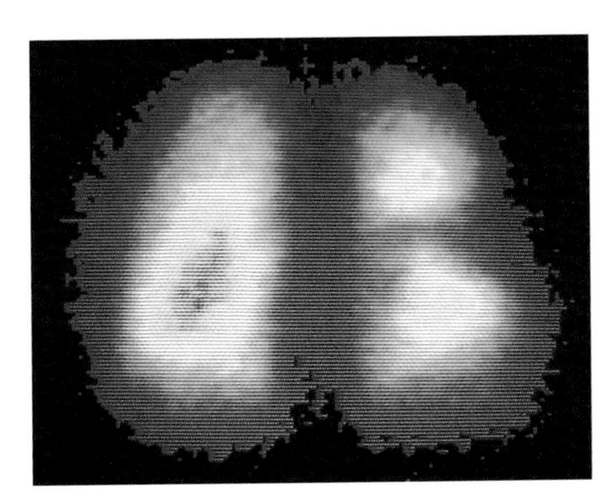

Figure 7.6 Scan of a lung with a blood clot (embolism). The lung on the right shows a restriction in its blood supply (blue 'notch') compared with the normal lung on the left

Lung function

Technetium-99m may be used to label coagulated albumen (a component of blood plasma). This is then injected into the bloodstream where it travels round the body. The coagulated albumen is trapped in the very fine blood capillaries in the lungs. Scanners can then be used to assess the efficiency of the blood flow through the lungs. If part of a lung is blocked by a blood clot, then the affected part of the lung will show reduced activity (see Figure 7.6).

Bone cancers

Bone cancers tend to produce extra bone growth on affected parts of the skeleton. Technetium-99m can be used to label phosphate molecules. If a cancer is present, there will be extra uptake of the phosphate at that point. The increased uptake can then be detected using, for example, a gamma camera (see Chapter 8). An example of a scan produced is shown in Figure 7.7.

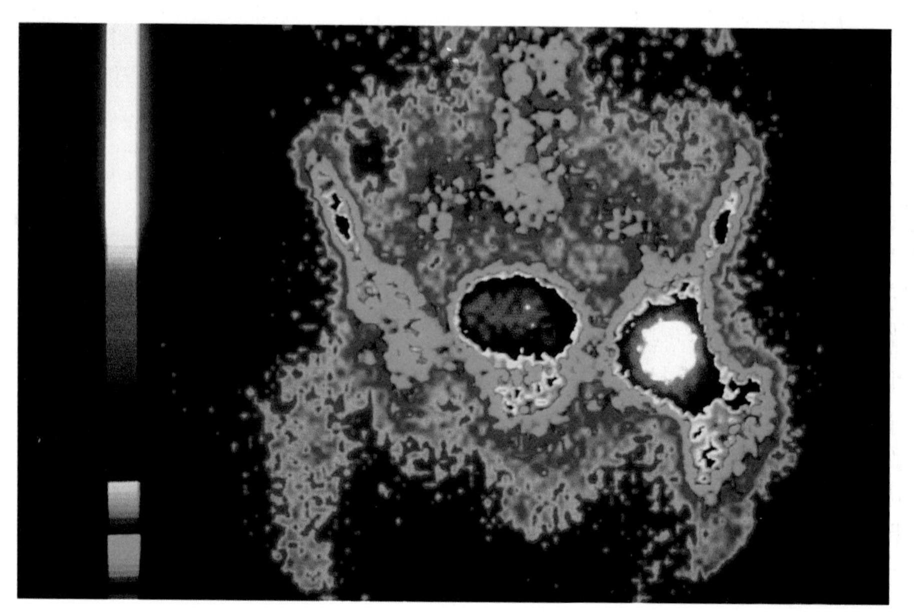

Figure 7.7 Gamma camera scan showing a bone cancer in the ball of the hip joint (the white 'hot spot' on the right)

A summary of the radioisotopes used in imaging is given in Table 7.3.

Table 7.3 Some radioisotopes used in medical imaging

Radioisotope		Preparation	Use
Technetium-99m	$^{99}Tc^{m}$	Coagulated albumen Sodium pertechnate Phosphate molecule Red blood cells	Blood flow in lungs Blood flow in brain Bone growth Heart function, blood circulation
Iodine-123 Iodine-131	^{123}I ^{131}I	Iodine Hippuran	Thyroid function Kidney function
Xenon-133	^{133}Xe	Gas	Lung function

Body fluids

A calculation involving the assessment of blood volume using iodine-131 is given in question 4 on page 108. The radioisotope tritium (^{3}H) may be used to label water molecules (tritiated water). A similar technique to that for the assessment of blood volume is then used to measure total body water volume.

2 Therapy

The use of iodine-131 or iodine-123 in the treatment of the thyroid has already been mentioned. An overactive gland may be treated with a moderate dose of iodine having an initial activity of about 300 MBq. Higher activities (about 2000 MBq) are used to treat thyroid cancers.

For some cancers (for example, cancer of the uterus), a sealed radioactive source may be inserted into the cancer itself. The intention is that the radiation emitted should kill the cancerous cells. Since cancerous cells divide more rapidly than normal cells and radiation damage is greater in cells that are dividing, the cancerous cells are killed at a greater rate than healthy cells. The isotopes used are usually beta emitters (so that the radiation dose is localised) with half-lives of a few days.

Radiotherapy is more commonly carried out using external sources of X-rays, gamma rays from cobalt-60, or electron beams. X-ray beams and electron beams are used in preference to gamma radiation, since the energy of the X-ray photons or electrons can be varied and controlled so as to produce maximum damage within the cancer.

Questions

1 Use the data in Table 7.1 on page 97 to answer this question.

a A sample of blood contains some iodine-131 and has an activity of 3.7×10^5 Bq. The activity was determined 36 hours after taking the sample. Calculate the activity of the sample when it was taken.

b A sample of air has a volume of 63 cm^3. The air contains radon gas with a total activity of 7.3 Bq. Determine the average number density of radon atoms in the air.

c A sample of phosphorus-32 is to have an activity of 3.2 MBq when it is used. The nuclide is prepared 120 hours in advance. What must be the activity of the freshly prepared sample?

2 Molybdenum-99 has a half-life of 67 hours. A sample of molybdenum, when first prepared, has an activity of 6.5×10^8 Bq. The Avogadro constant is 6.0×10^{23} mol^{-1}.

Determine, for the sample of molybdenum-99,

a the decay constant in s^{-1},

b the initial mass of the sample,

c the activity after 72 hours.

3 A radioisotope has a physical half-life of 14 days. A sample of the isotope is injected into a patient. The maximum activity within a gland is found to occur after 12 hours. The activity is 0.40 times this maximum activity 180 hours after first injecting the isotope.

Determine, for this isotope in this gland,

a its effective half-life,

b its biological half-life.

4 The biological and the physical half-lives of iodine-131 are 21 days and 8 days respectively.

In order to find the total blood volume of a patient, 5.0 cm^3 of a substance labelled with iodine-131 was injected into the bloodstream. After 20 minutes, a sample of blood having a volume of 5.0 cm^3 was taken from the patient. The sample had an activity of 90 Bq.

A second sample of the labelled substance having a volume of 10 cm^3 was diluted with 1000 cm^3 of water. A 5.0 cm^3 sample of this diluted mixture had an activity of 600 Bq.

a Determine:

i) the total activity of a 5.0 cm^3 sample of the labelled substance,

ii) the total volume of the patient's blood.

b Suggest why half-life is not taken into consideration when calculating blood volume.

c State one assumption made about the mixing of the blood.

5 Sodium-24 ($^{24}_{11}Na$) may be produced by bombarding sodium-23 with slow neutrons in a nuclear reactor.

a Write down a nuclear equation for this reaction.

b Sodium-24 is a beta emitter with a half-life of 15 hours. Suggest one advantage and one disadvantage of its use in radiotherapy.

6 A sample of molybdenum in a technetium-99m generator has a half-life of 67 hours and an initial activity of 6.0×10^9 Bq. Technetium-99m is extracted from the generator at 36-hour intervals. At each extraction, 90% of the activity is removed.

Calculate the number of extractions that can be made if the activity of the extracted sample is not to fall below 5% of the initial activity of the molybdenum.

8 Radiation detectors

In this chapter you will read about:

- the structure and operation of a film badge
- the principle of the thermoluminescent dosimeter
- the structure and operation of a scintillation counter
- the structure and operation of a gamma camera
- the structure and operation of a magnetic resonance imaging scanner
- the principle of fluoroscopy
- the principle of CT scanning

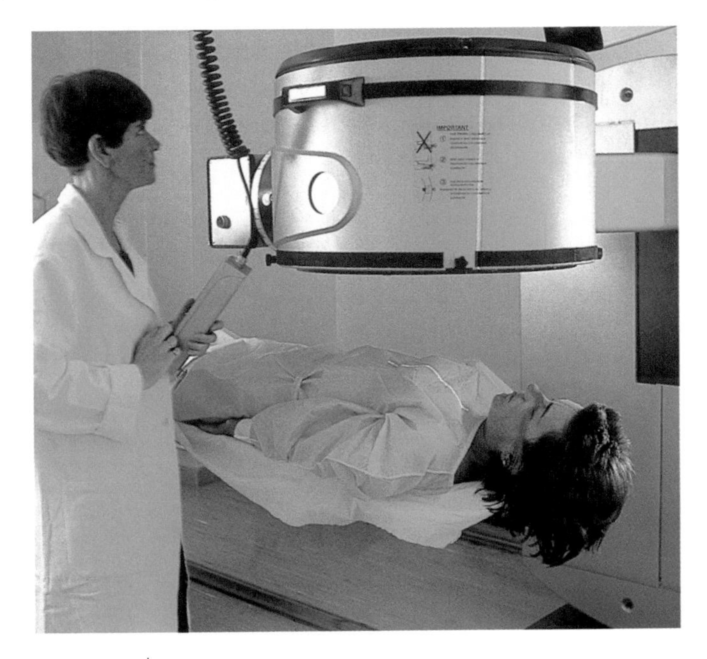

Figure 8.1 A gamma camera in use

Introduction

Most of the radiations that are detected for diagnostic purposes are electromagnetic in nature. Four commonly used detectors of ionising radiations and one detector of radio frequency photons of a lower energy will be described in this chapter. There is a wide range of frequencies and hence photon energies of radiation involved. The detection mechanisms rely on the interaction of these radiations with certain atoms. The effect of the interaction is determined by the energy of the incident photons of radiation. There is no single atom (and hence no single detector) that interacts with the whole range of photon energies. It is therefore necessary for a number of different detectors to cover and detect radiation from the whole range of the spectrum of ionising radiations.

The film badge

Devices that monitor the dose received (see Chapter 5) by an individual are called **dosimeters**. The film badge is such a device. It records personal exposure to radiation and allows an estimate to be made of the total dose received by the wearer. The accuracy of the film badge is low. A film badge typically detects only between 10 and 20% of the incident radiation, but this is sufficient for low-level monitoring and recording of the dose received.

The effect of ionising radiation on photographic film

Ionising radiation has the same effect as light on undeveloped photographic film. Black and white photographic film has a cellulose base covered in an emulsion that contains grains of silver bromide suspended in gelatin. When ionising radiation is incident on the film, a **latent image** is formed as some of the silver ions turn into silver atoms, which are black (Figure 8.2a).

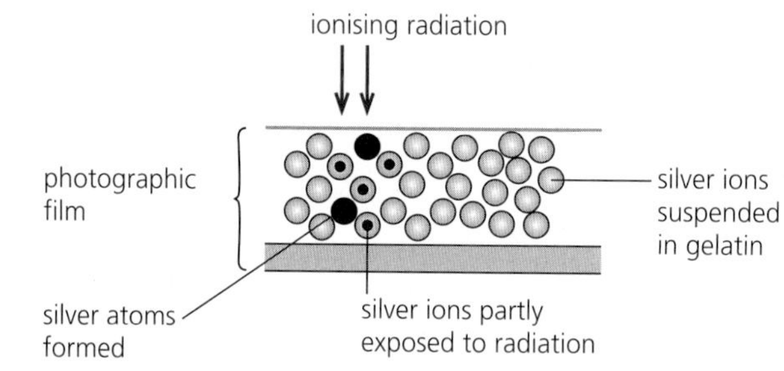

(a) The effect of ionising radiation on photographic film

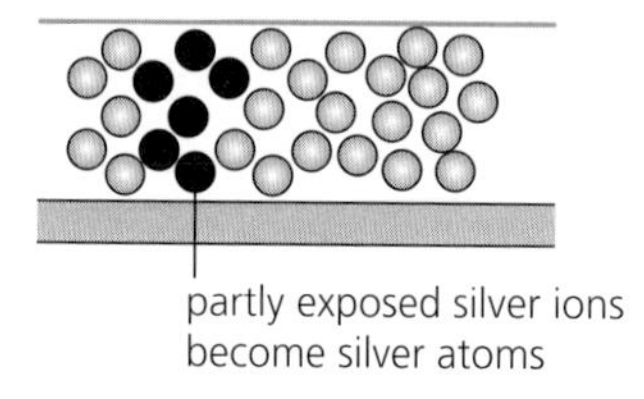

(b) Developing the film

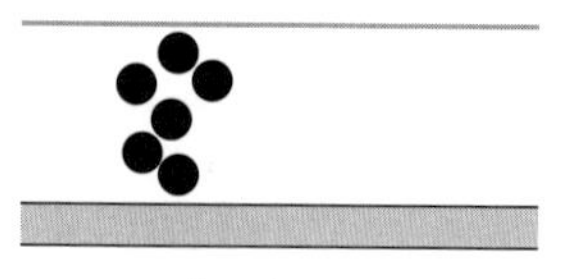

(c) Fixing the film

Figure 8.2

WORKED EXAMPLE 8.1

A photomultiplier tube has nine dynodes made of a material that releases four secondary electrons for each incident electron on its surface. Calculate, for a single electron incident on the first dynode:

a the number of secondary electrons emitted from the first dynode,
b the number of secondary electrons emitted from the second dynode,
c the number of secondary electrons emitted from the ninth dynode,
d the gain of the photomultiplier tube.

a One electron landing on the surface ejects 4 electrons.

b Each of these electrons ejects 4 more electrons from the second surface.

$$4 \times 4 = 16 \text{ electrons}$$

c $4^9 = 2.6 \times 10^5$ electrons

d $$\text{Gain} = \frac{\text{output}}{\text{input}}$$

$$= \frac{2.6 \times 10^5}{1}$$

$$= 2.6 \times 10^5$$

The collimator

The direction of emission of a scintillation caused within a scintillation crystal is independent of the direction of travel of the incident radiation. This means that stray radiations from areas not being investigated may cause unwanted scintillations (see Figure 8.6a overleaf).

To reduce the amount of stray radiation entering the scintillation counter, a collimator, made of lead, is attached to the end of the counter. Only radiations that enter along a path parallel to the cylindrical axis of the collimator strike the crystal at the end and possibly result in a scintillation (Figure 8.6b). This results in a sharper image that also contains a superimposed grid due to the collimator.

Figure 8.7 shows a variation in the head of the scintillation counter which enables the activity of liquid samples to be measured. The crystal surrounds the radioactive sample and thus a greater proportion of the emitted radiation ends up in the crystal.

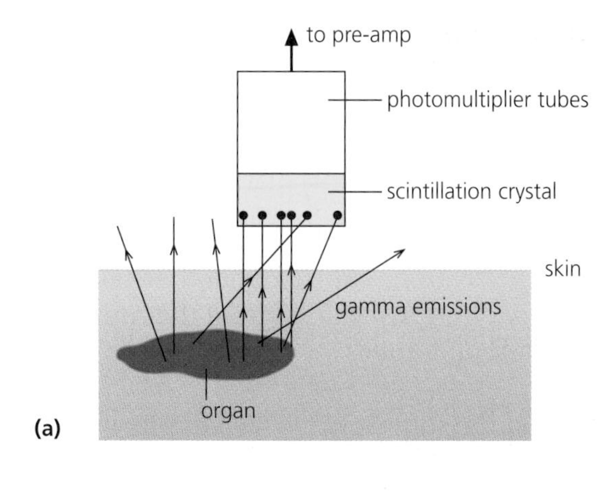

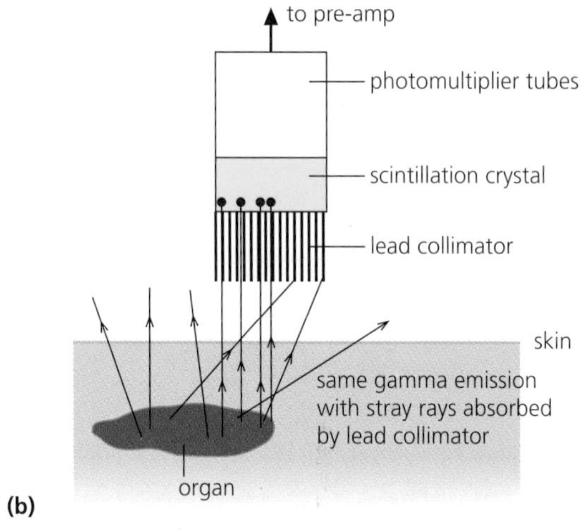

Figure 8.6 The action of a collimator

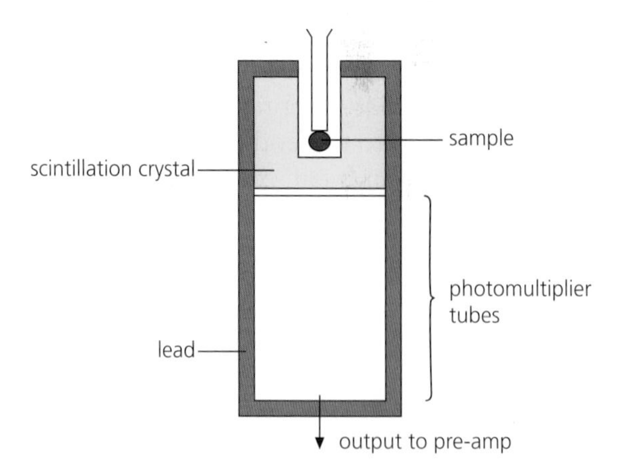

Figure 8.7 A 'well-type' scintillator head

Use of the scintillation counter

The scintillation counter, with the attachment of the appropriate scintillation crystal, is able to detect and make estimates of the activity of the source of the following radiations: alpha, beta, gamma, X- and neutron radiations.

Radiation may be monitored by the scintillation counter in 'tracer' studies. A radionuclide is prepared and a measured quantity is given to the patient, either orally in a solution or by injection into a blood vessel, depending on the part of the body under investigation. Time is allowed for the radionuclide to mix with the surrounding medium. The emissions of radiation from the decaying radionuclide are then detected outside the body by the scintillation counter.

Blockages in vessels may be located and regions in which there is restricted blood flow may be identified, such as in the determination of the part of the brain affected in a patient who has had a stroke.

Another common use for monitoring radiation emitted from the body is in the study of thyroid uptake – see Chapter 7, pages 100 and 107.

The rectilinear scanner

The scintillation counter provides discrete information about radiation emissions from a very small area of a patient. In order to build up a picture of the activity from within a region of a patient, it is necessary to move the scintillation counter, take a reading and move the counter on again, recording the readings taken at each position. The **rectilinear scanner** utilises the scintillation counter in this way. In order to make sense of the readings taken at each position, a photograph is produced by making the brightness of the image on the film correspond to the intensity of the radiation received at the scintillation counter. Figure 8.8 shows the principle of a rectilinear scanner

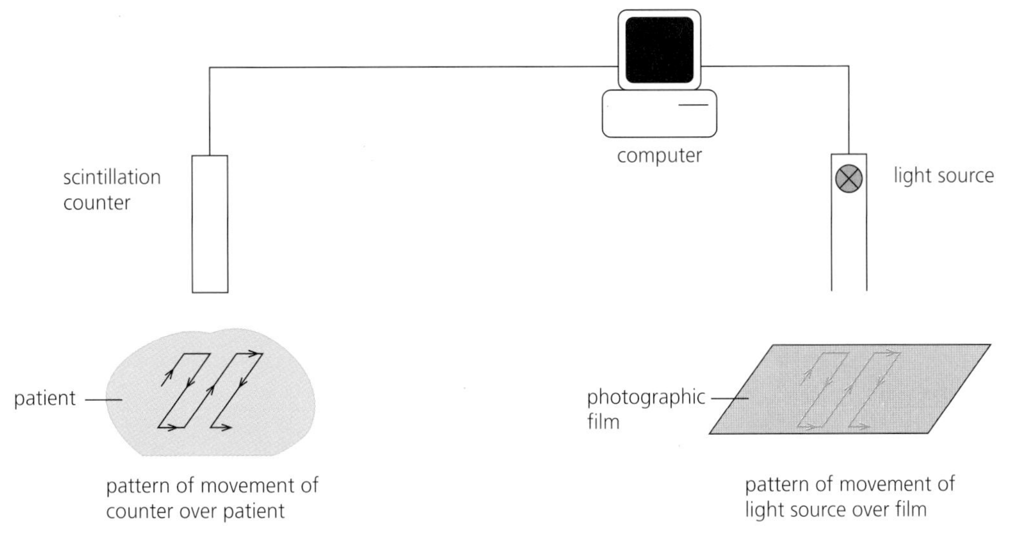

Figure 8.8 The rectilinear scanner

A computer controls the movement of the scintillation counter over a patient. A light source is moved over a photographic film, replicating exactly the movement of the scintillation counter. The intensity of the light source is controlled by the computer and is proportional to the signal from the scintillation counter. When developed, the film shows an image of the source of radiation from within the body.

The rectilinear scanner has now mostly been superseded by the gamma camera (see below). Both devices monitor the gamma emissions from radioisotopes that have been mixed with the blood which flows through a patient.

The gamma camera

The gamma camera (see Figure 8.1, page 109) works on the same basic principle as the scintillation counter. It is in essence many scintillation counters grouped together. The output from each counter is used as part of the data required to form an image. Figure 8.9 shows a schematic diagram of a basic gamma camera.

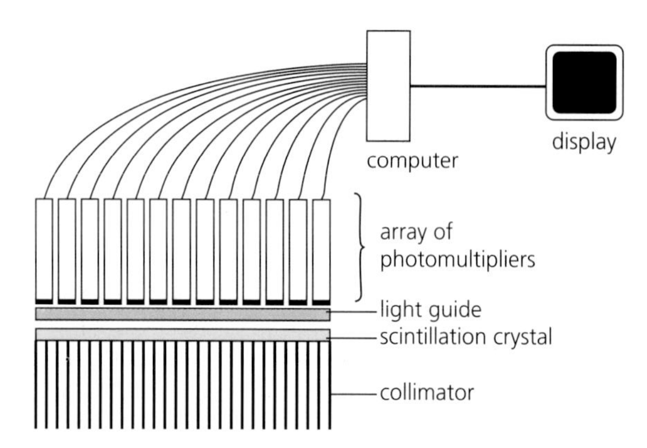

Figure 8.9 Schematic diagram of a gamma camera

The collimator comprises a honeycomb of open-ended cylindrical lead tubes. These lead to one single large scintillation crystal, above which is an array of 20 or more photomultiplier tubes. The output from each photomultiplier tube is amplified and processed by a computer. The intensity of the pulses in each tube is used to build up a precise location of the source since tubes further from the scintillation receive a smaller intensity light flash.

Any gamma photons that are not parallel to the collimator walls are absorbed by the lead walls and do not progress to the crystal. Between 20 and 25% of the gamma photons incident on the crystal produce scintillations. These light photons are emitted in all directions. Some will end up landing perpendicular to the plane of the photocathode and will contribute to the imaging process. Other (stray) light photons will be incident at angles other than 90° to the plane of the photocathode (Figure 8.10). These can cause the image formed to be less clear.

An example of an image produced by a gamma camera is given in Figure 7.7 (page 106).

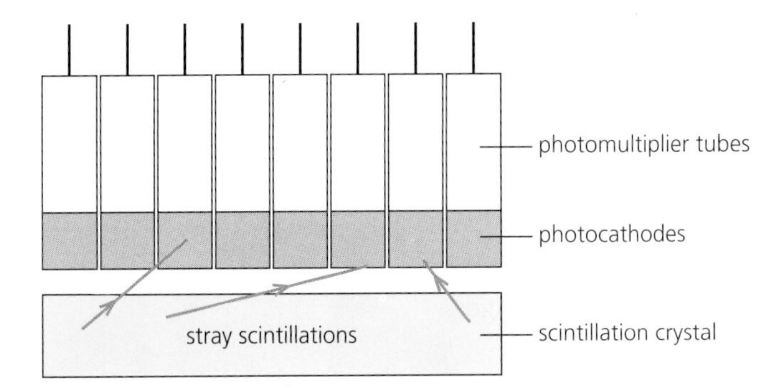

Figure 8.10 Stray (non-perpendicular) photons, which can cause blurring of the image

Magnetic resonance imaging

Magnetic resonance imaging (MRI) provides the most detail of all the imaging processes currently used in medicine. The radiation used is in the radio frequency region of the electromagnetic spectrum and exposure for a scan is considered to carry no risk at all. The scanning process is non-invasive and, apart from a noisy and claustrophobic environment during the scan, a patient experiences no discomfort. Figure 8.11 shows a patient undergoing an MRI scan and the resultant image of a section through the head.

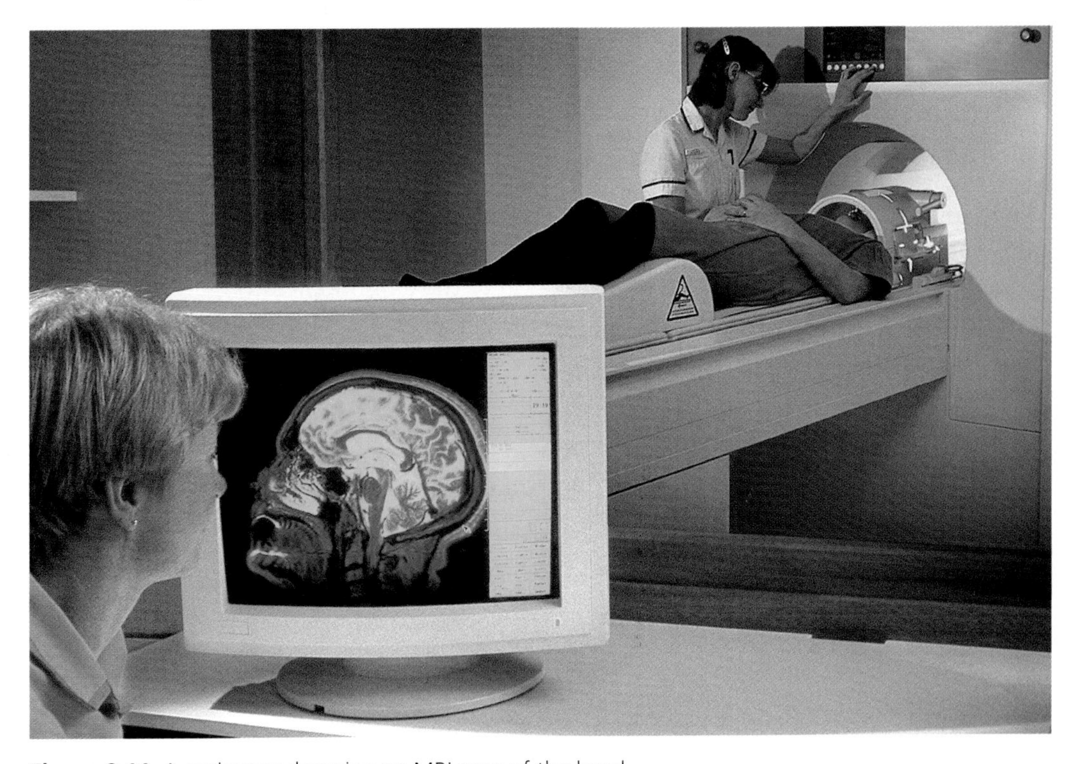

Figure 8.11 A patient undergoing an MRI scan of the head

An MRI image is formed after the compilation of data from radio frequency radiation emitted by certain atoms within the body after they have been made to resonate. The intensity of radio frequency radiation detected is proportional to the number of atoms of a given type that are present. It is therefore possible to plot the relative distributions of these atoms and hence produce a picture of the structure being scanned.

The hydrogen atom exhibits all of the properties required for magnetic resonance imaging. It is one of the most abundant atoms in the human body and so it is most frequently used to generate an image.

Nuclear magnetic resonance

Some atoms (such as hydrogen) have an unequal number of protons and neutrons in the nucleus. When these atoms spin, regions of charge within the nucleus move. Moving charges have, associated with them, a magnetic field. The nuclei therefore act like tiny magnets (Figure 8.12).

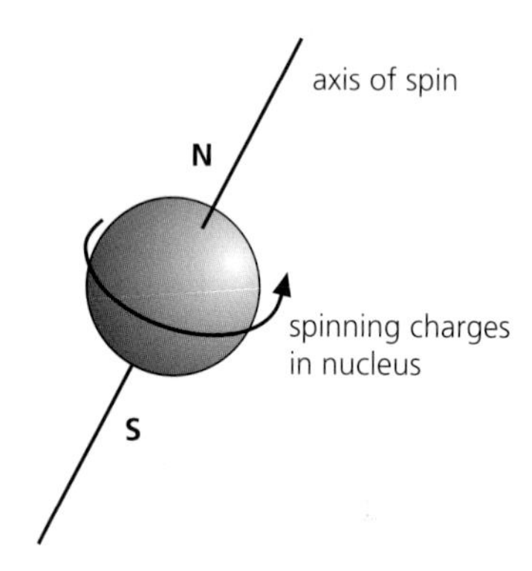

Figure 8.12 The magnetic poles of a spinning nucleus

When an external magnetic field is applied to the atoms, they tend to align themselves in the field. The alignment of the axis of spin is not exact, however. The axis of spin performs a circular rotation about the direction of the external field (Figure 8.13). The spinning atom is said to be **precessing** in the external magnetic field.

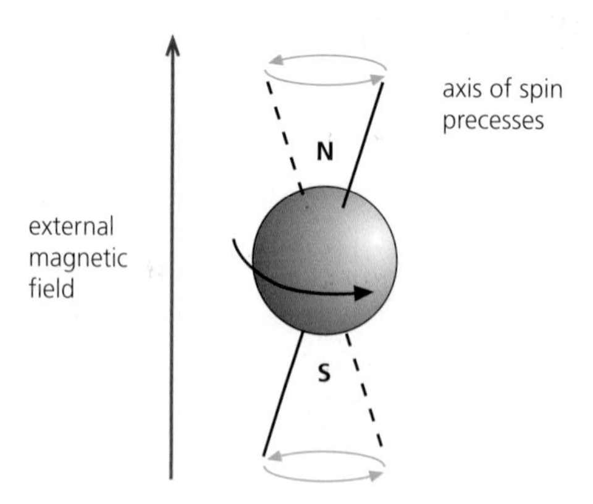

Figure 8.13 Precession in an applied magnetic field

The frequency of precession (called the Larmor frequency) for a given atom is determined by the strength of the external magnetic field as well as by the structure of the nucleus. The motion of a precessing atom causes a tiny electromagnetic signal at the Larmor frequency.

In MRI an external pulse of electromagnetic radiation is applied to the patient at the Larmor frequency for hydrogen atoms. The precessing atoms absorb this radiation and resonance occurs. The amplitude of precession builds for the duration of the pulse. When the pulse stops, the amplitude of precession dies down as the atom returns to the unexcited state. This takes a short period of time (called the relaxation period), which may be measured by monitoring the decreasing amplitude of the emitted radio frequency signal which induces an e.m.f. across a coil. This information is stored in digital form and processed by a computer. Two relaxation processes occur, producing two relaxation times. The combination of these times is used to produce an image.

The MRI scanner

Figure 8.14 shows a cut-away diagram of an MRI scanner.

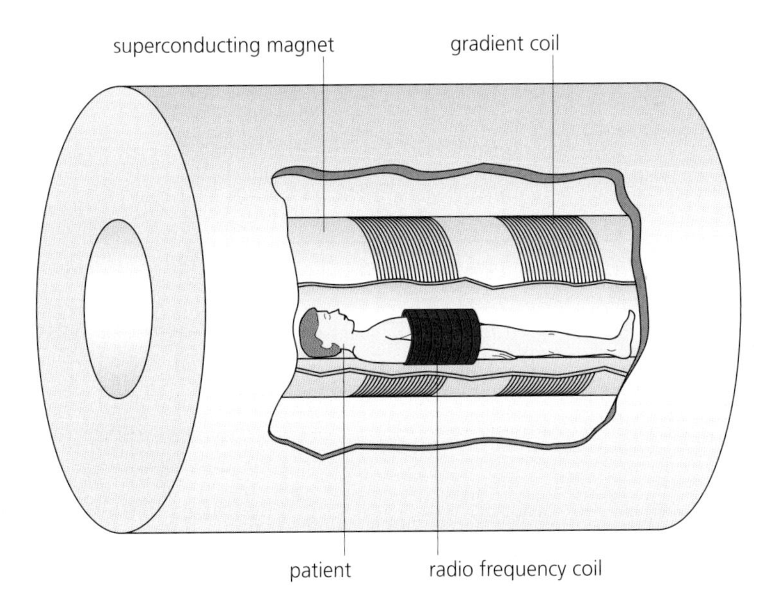

Figure 8.14 An MRI scanner

At the heart of the scanner is the large fixed magnet with a magnetic flux density of about 2 tesla. In order to produce such a strong magnetic field, superconducting materials are used. These materials have no resistance to the flow of electric current when they are at very low temperatures, typically below −269 °C. To achieve these temperatures, the materials are immersed in liquid helium. Part of the running expense for the hospital is the replacement of the liquid helium which boils away.

In order to locate the source of the radio frequency signals, there are usually three separate **magnetic field gradient** coils, arranged in such a way so as to add a small and gradually increasing field across the length, width and depth of the patient. This assigns a specific magnitude of the magnetic flux density at each point within a patient. As the frequency of precession is determined in part by the magnetic flux density, the signal from each part of the patient has a slightly different Larmor frequency. The computer is therefore able to determine the position of the source of each signal and to use this information to construct an image.

The MRI scanner is expensive both in the initial capital cost and in the day-to-day running, but the quality of the images produced and the detailed information gained is superior to other imaging processes. This imaging technique is currently used principally to image the brain and the central nervous system.

Fluoroscopy

X-ray **fluoroscopy** is a process whereby a continuous X-ray beam is passed through a patient and onto a fluorescent screen where a dim, real-time image is formed. In order to view the image directly and gain useful information, it would be necessary for the patient to be exposed to unacceptably high levels of X-radiation. This problem is partly overcome by employing an **image intensifier** to amplify the output intensity at each point on the screen (Figure 8.15).

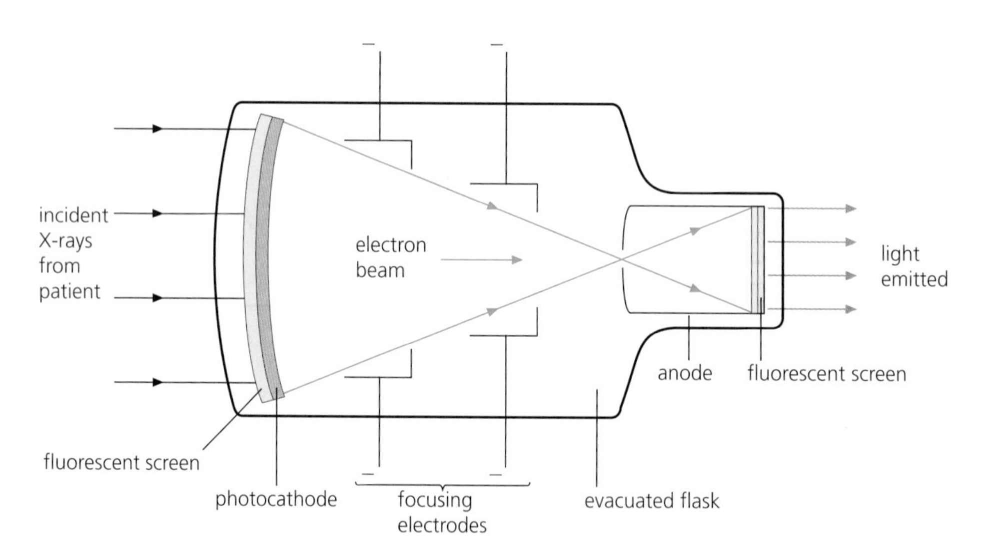

Figure 8.15 An image intensifier

After passing through a patient, X-rays strike the first fluorescent screen. Light photons are emitted at the screen, and release electrons through the photoelectric effect at the photocathode. A potential difference of about 20 kV accelerates the electrons across an evacuated glass tube where they are focused by a series of electrodes onto a smaller fluorescent screen. The intensity of light output at this second screen is increased due to (i) the increased rate of electrons arriving per unit area, and (ii) the increased energy of the electrons. The image formed is sent to a television screen via a TV camera.

The information gained allows images of a moving patient as opposed to the static image produced by a short burst of X-radiation. Although use of the image intensifier in X-ray fluoroscopy reduces the X-ray dose received by a patient, this technique still involves a greater exposure to X-rays than other radiographic imaging processes (due to the continuous beam of X-rays) and consequently it is used sparingly.

Intensifying screens

Intensifying screens are used in conventional X-ray imaging to reduce the time required for X-rays to affect photographic film. A double-sided photographic film is sandwiched between two intensifying screens (Figure 8.16).

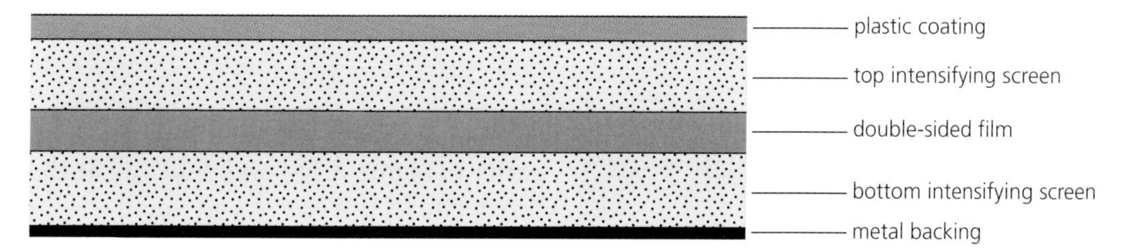

Figure 8.16 The use of intensifying screens

The intensifying screens contain fluorescent crystals (usually zinc sulphide). The atoms of these crystals absorb X-ray photons, become excited and return to the equilibrium state emitting photons of light. Light is emitted in all directions but those light photons that travel directly towards the photographic film cause blackening of the film. The thin intensifying screen is in direct contact with the photographic film, so the number of light photons that diverge prior to hitting the film is minimised and the image is sharp.

The film is more sensitive to light photons than to X-ray photons. The production of light photons therefore speeds up the imaging process and reduces the required X-ray exposure time by a factor of as much as 200.

Computer tomography (CT) scanner

A **CT scanner** produces two-dimensional images of sections through a patient (Figure 8.17a). If these sections are taken in order, along the length of a patient, a three-dimensional image may be constructed by a computer (Figure 8.17b).

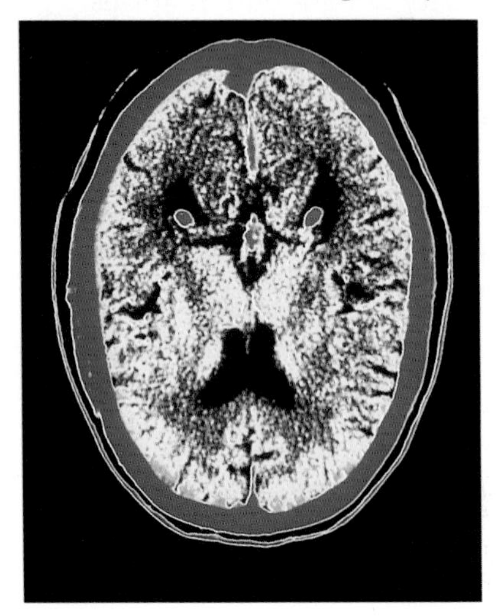

(a) Two-dimensional CT scan of the brain showing the cerebral hemispheres

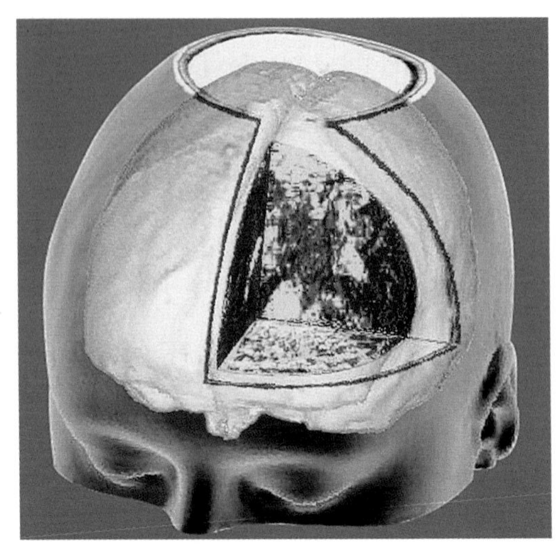

(b) Three-dimensional CT image constructed from two-dimensional scans

Figure 8.17

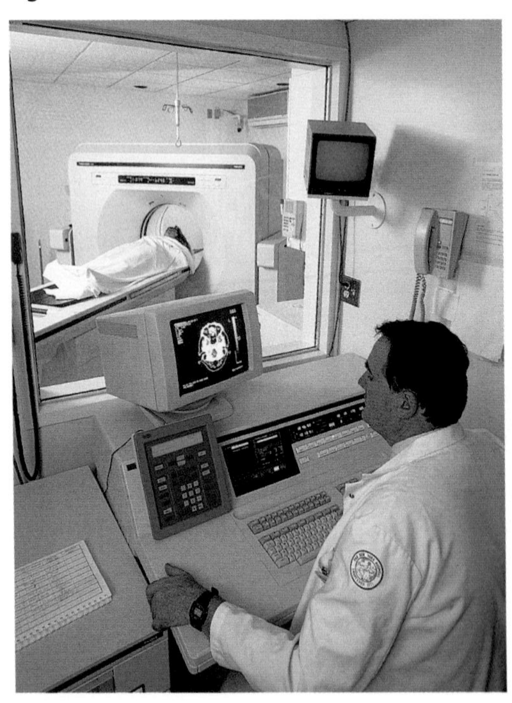

Figure 8.18 A CT scanner in operation

An X-ray beam is pulsed through a patient (Figure 8.18) and detected at the other side by a series of very small detectors. The X-ray beam and detectors are mounted on a gantry.

The signal received by each detector is recorded by the computer and stored. The gantry is then rotated about the patient through a small angle, the X-ray beam pulsed once again and the resulting information stored by the computer. This process is repeated periodically for angles up to 180°. The computer divides the body up into small unit volumes called 'cells'. The intensity of the X-ray beam that passes through a given cell from different directions is recorded and, by a series of simultaneous equations, the attenuation of the X-ray beam by a given cell may be calculated.

Consider nine cells in a section of soft tissue (Figure 8.19). The numbers in each cell show the percentage of the incident X-ray pulse attenuated by that cell due to the medium of the cell. As an X-ray beam is pulsed horizontally through the top three cells, the detector records that 12/100 of the incident beam intensity has been attenuated. The beam moves to a new position, pulsing horizontally through the second row of cells, and the process of recording the fraction of the beam that has been attenuated is repeated. For the second row the fraction is 8/100.

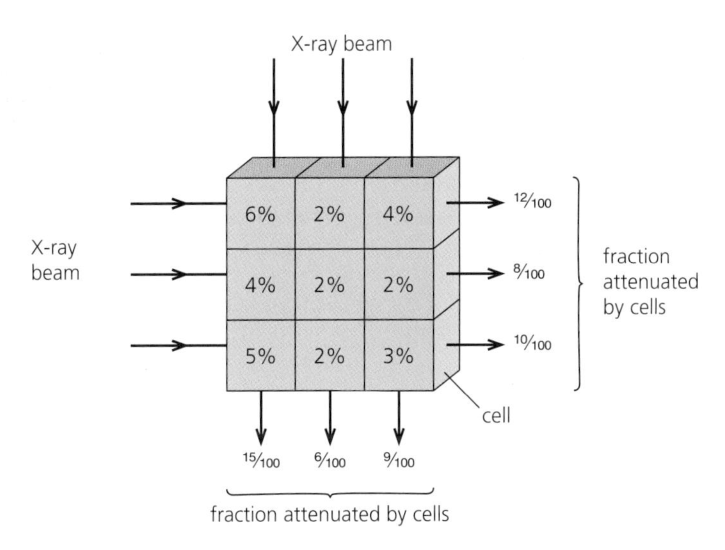

Figure 8.19 Attenuation of the X-ray beam in two directions

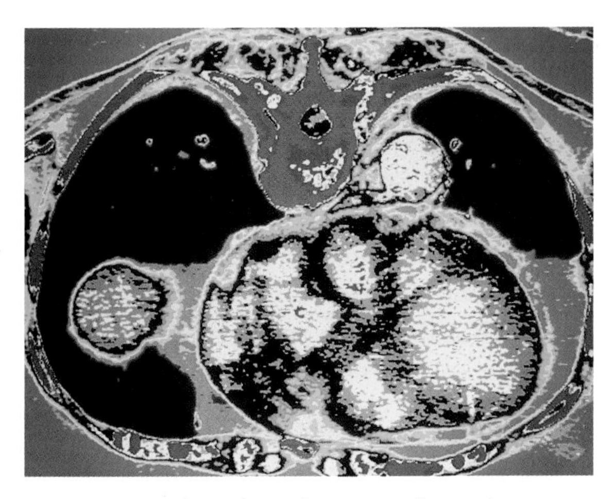

Figure 8.20 False-coloured CT scan of a section through the chest. The heart and major blood vessels are shown here in yellow, and the bones in pink

When all of the data has been collected, the computer calculates through simultaneous equations the attenuation contributed by each cell. This information is then converted into a picture by assigning a greater degree of darkening to a cell with a greater attenuation value. An image is formed where darker regions correspond to media such as bone, and lighter regions correspond to softer tissue such as muscle. Figure 8.20 shows a CT scan of a section through the chest.

Questions

1 a Explain the function of:

i) the scintillator,

ii) the photocathode, and

iii) the dynodes

in a scintillation counter.

b The anode current in a scintillation counter is 6.5×10^{-14} A, when beta particles are incident on the scintillation crystal. Each of the nine dynodes is made of the same material. For each electron incident on a dynode, four secondary electrons are released. ($e = 1.6 \times 10^{-19}$ C)

i) Calculate the rate of arrival of electrons at the anode.

ii) Calculate the rate of arrival of photoelectrons at the first dynode.

iii) Calculate the number of light photons arriving at the photocathode each second if the photocathode emits one photoelectron for every five incident light photons.

iv) Calculate the number of beta particles detected each second if the scintillation crystal converts 10% of the incident beta particle energy into light photons.

2 Figure 8.21 shows a diagram of a film badge alongside a developed film.

a State, giving reasons, the nature of the radiations responsible for the blackening of the film shown.

b Explain how the film badge detects:

i) thermal neutrons,

ii) very high-energy X-rays.

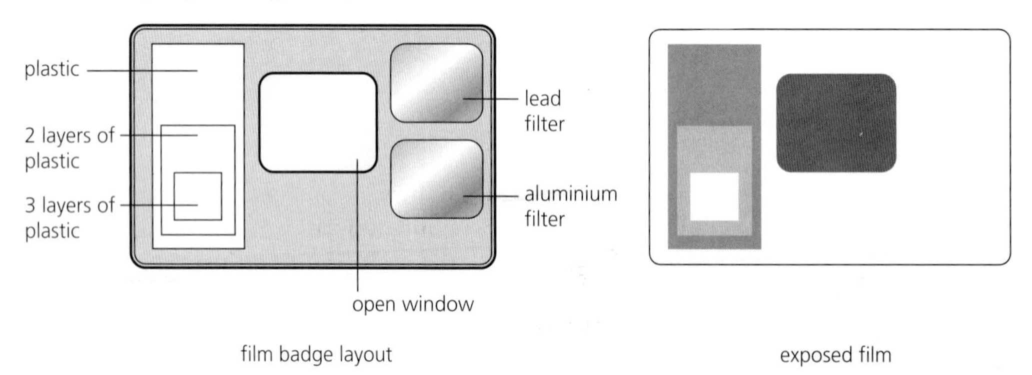

Figure 8.21

3 a Describe the principle of operation of the thermoluminescent dosimeter.

b Explain three advantages and one disadvantage of the thermoluminescent dosimeter over the film badge.

4 a Explain how information from within a patient is detected by a gamma camera and converted into an image.

b Explain briefly the physics principles employed in the use of a rectilinear scanner.

c Compare, from a patient's point of view, the process of being imaged by each of the above methods.

5 a Describe the use of X-ray fluoroscopy, comparing the form of the information gained with that from a standard X-ray image.

b The area of the large fluorescent screen of an image intensifier is $0.25\,m^2$. The area of the small fluorescent screen is $2.5 \times 10^{-4}\,m^2$. Calculate the fractional increase in intensity of the image at the small screen due to the *structure* of the image intensifier. (For this part of the question you are to ignore the increase in intensity due to the extra energy gained by the electrons as a consequence of the p.d. across the image intensifier.)

Ultrasound

In this chapter you will read about:

- the nature of ultrasound
- the generation and detection of ultrasound
- the factors that affect the reflection of ultrasound
- how ultrasound is used to obtain diagnostic information

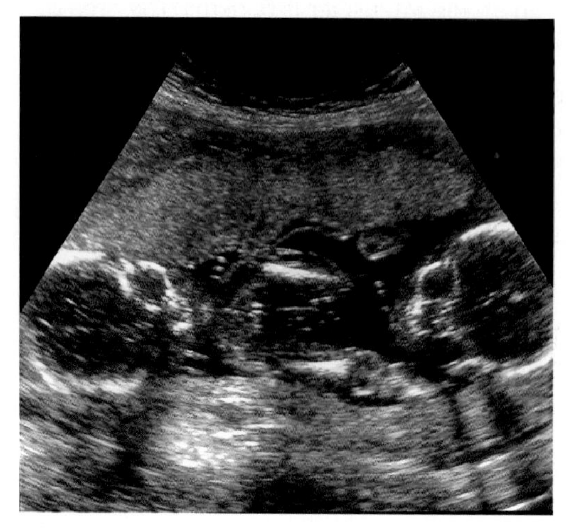

Figure 9.1 Ultrasound scan of twin foetuses

Introduction

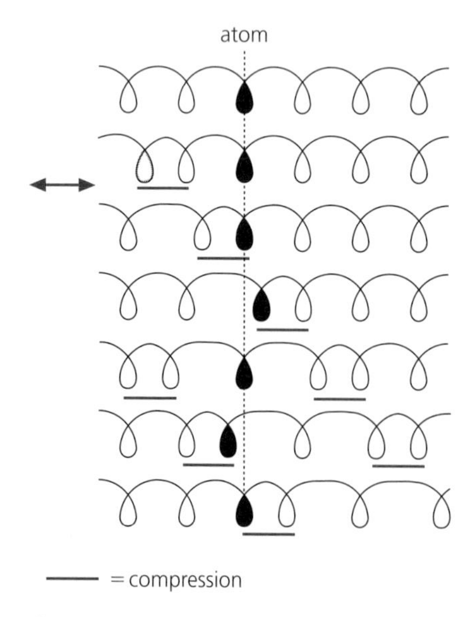

Figure 9.2 A compression wave

Sound is a compression wave in which energy is transferred as a series of collisions between the oscillating atoms or molecules of the substance (medium) through which the sound is travelling. Figure 9.2 shows a macroscopic model of a compression wave on a spring. The black coil represents an atom (or molecule) of the medium through which the compression wave is travelling. The displacement of this atom (or molecule) is shown at given times as the wave passes through.

The amplitude of vibration of these atoms or molecules increases and then decreases in the direction of propagation of the sound wave as the wave passes through. The frequency of the compressions that result determines the pitch of the sound. The human ear may detect the frequencies of these compressions between 20 Hz and 20 000 Hz (20 kHz) (see Chapter 2). Frequencies above 20 kHz are known as **ultrasonic** frequencies and these waves are referred to as **ultrasound**. When ultrasound is used for imaging purposes, the frequencies required are in the 1 MHz–10 MHz region (1000–10 000 kHz).

Generation of ultrasound

In order to generate an ultrasound wave, the atoms of a dense medium are required to oscillate with a large amplitude and at a frequency above 20 kHz. The dense medium employed for the purpose of generating ultrasound is made from a **piezoelectric** crystal. When a potential difference is applied across a piezoelectric crystal, it changes shape. An example of a piezoelectric crystal is quartz, whose structure is made up of repeating tetrahedral silicate units. One such unit is shown in Figure 9.3a.

The crystal lattice is sandwiched between two layers of silver that act as electrodes. When a potential difference is applied across the layers of silver, the crystal deforms in the electric field. When the lower layer of silver is at a positive potential it will attract the negative oxygen ions of the silicate molecule (Figure 9.3b). The upper layer of silver is at a negative potential, so it will attract the positive silicon ions. The structure is thus stretched.

Reversing the p.d. across the silver electrodes has the effect of compressing the crystal (Figure 9.3c). The lower layer of silver is at a negative potential, so it repels the negative oxygen ions while the upper positive silver electrode repels the silicon ions.

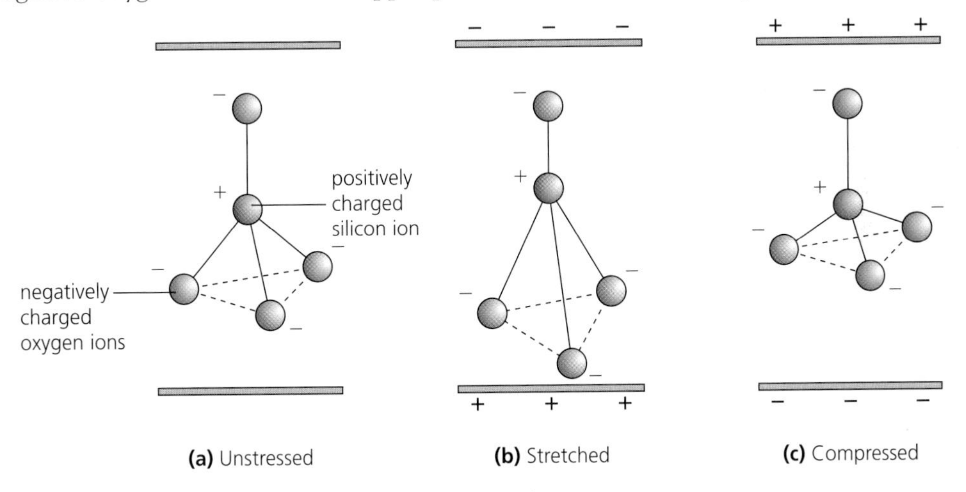

Figure 9.3 A silicate molecule in a piezoelectric crystal

Piezoelectric crystals such as quartz or lead zirconate titanate (PZT) are made to vibrate at their resonant frequency in order to maximise the amplitude of vibration necessary for the production of ultrasound. This is achieved by applying an alternating p.d. across the crystal at the resonant frequency of the crystal. This is the frequency of the ultrasound that is produced. The piezoelectric crystal and electrodes are referred to as an **ultrasound transducer**. Figure 9.4 shows an ultrasound transducer in use for imaging purposes.

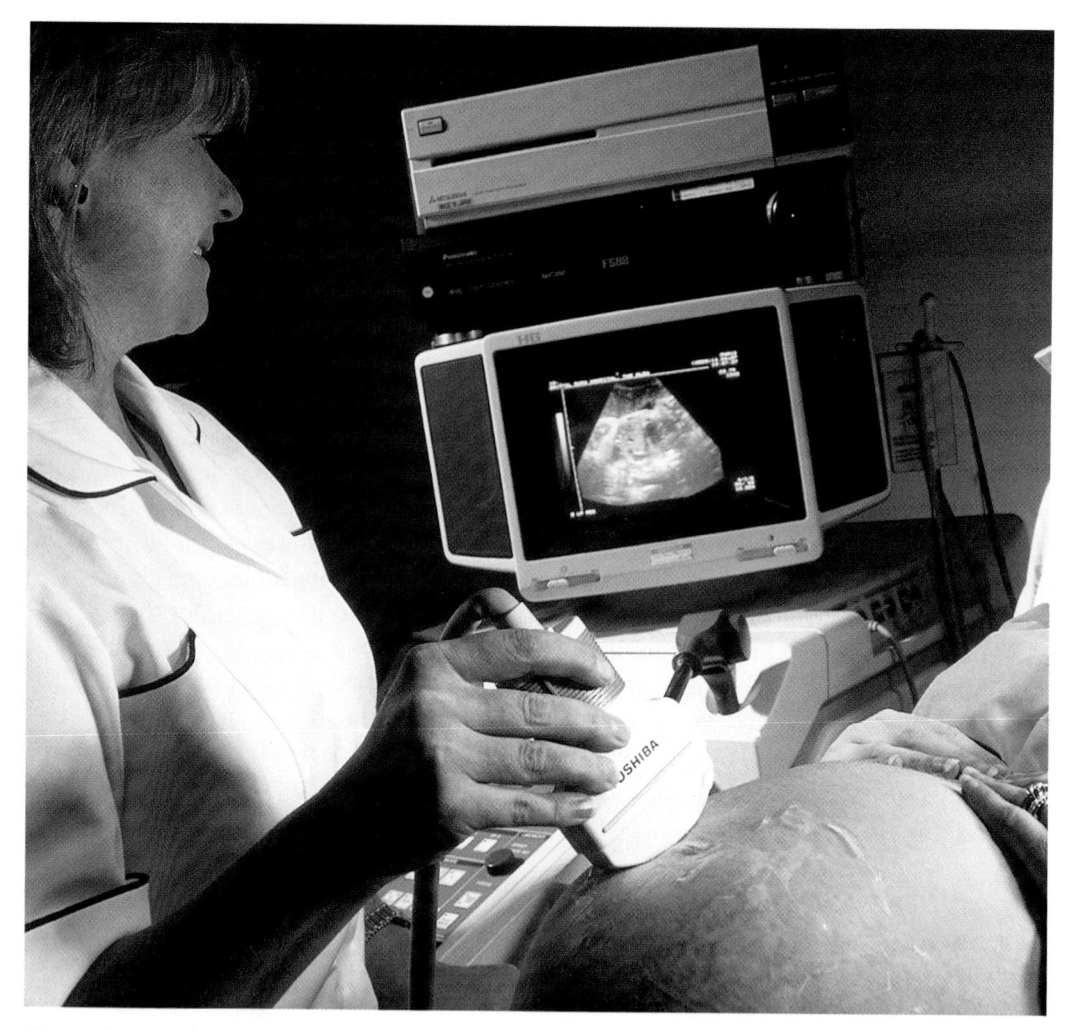

Figure 9.4 An ultrasound transducer being used to obtain an image of a foetus

Detection of ultrasound

When an ultrasound pulse is incident on a piezoelectric crystal, the crystal vibrates and an alternating p.d. is generated across the crystal. This happens as the oscillation of the crystal alters the positions of the positive and negative ions in the structure. The movement of these ions induces opposing charges on the electrodes and so a p.d. is set up between the electrodes. The same crystal may therefore be used both to generate and to detect an ultrasound pulse. Problems arise, however, if a pulse of ultrasound is received at the crystal before the crystal has stopped vibrating after generating a pulse. To overcome this, the alternating p.d. is pulsed for a short period of time, usually about 1 μs. In order to stop the crystal from vibrating after the alternating p.d. has ended, a backing material, such as an epoxy resin, damps the vibrating crystal (see Figure 9.5). The crystal is then ready to detect an incoming ultrasound pulse.

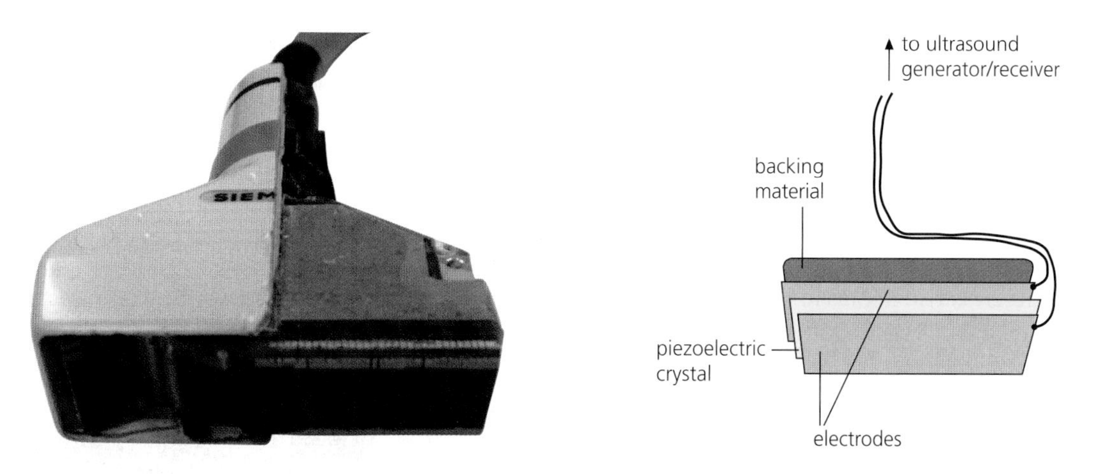

Figure 9.5 Construction of an ultrasound transducer

Reflection of ultrasound waves

An ultrasound wave is a pressure wave that may be reflected or refracted. Where ultrasound is incident on a boundary between two different substances (media), a fraction is reflected and the remainder is refracted (Figure 9.6).

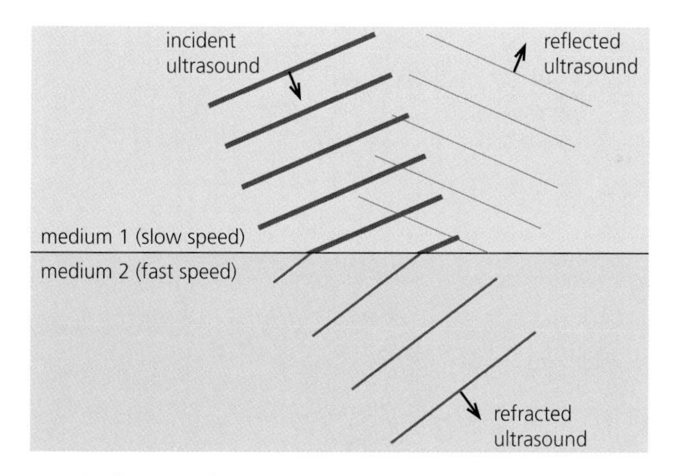

Figure 9.6 Reflection and refraction of ultrasound waves at a boundary between media

The fraction of the ultrasound reflected depends upon a property known as the **acoustic impedance** (Z) of the media on either side of the boundary. Acoustic impedance Z is the product of the density ρ of the medium and the speed c of ultrasound in the medium:

$$Z = \rho c$$

The unit of Z is $\mathrm{kg\,m^{-2}\,s^{-1}}$.

The fraction of the intensity of the incident ultrasound at a boundary that is reflected is called the **intensity reflection coefficient** (α). The value of α is given by the equation:

$$\alpha = \frac{(Z_2 - Z_1)^2}{(Z_2 + Z_1)^2}$$

where Z_1 is the acoustic impedance of the medium from which the ultrasound travels and Z_2 is the acoustic impedance of the medium into which the ultrasound travels.

Table 9.1 opposite gives data of the acoustic impedances of a variety of body tissues.

WORKED EXAMPLE 9.1

Explain, with reference to the following data, the need for the use of a gel between an ultrasound transducer and the skin. Begin your answer by calculating

a the fraction of ultrasound reflected at an air–skin boundary,
b the fraction of ultrasound reflected at a gel–skin boundary.

The values of the acoustic impedances for air, skin and gel are $4.29 \times 10^2\,\text{kg m}^{-2}\,\text{s}^{-1}$, $1.7 \times 10^6\,\text{kg m}^{-2}\,\text{s}^{-1}$ and $1.65 \times 10^6\,\text{kg m}^{-2}\,\text{s}^{-1}$ respectively.

a The acoustic impedance for air is $Z_1 = 4.29 \times 10^2\,\text{kg m}^{-2}\,\text{s}^{-1}$.
The acoustic impedance for skin is $Z_2 = 1.7 \times 10^6\,\text{kg m}^{-2}\,\text{s}^{-1}$.

$$\text{Fraction of ultrasound reflected} = \frac{(Z_2 - Z_1)^2}{(Z_2 + Z_1)^2}$$

$$= \frac{(1.7 \times 10^6 - 4.29 \times 10^2)^2}{(1.7 \times 10^6 + 4.29 \times 10^2)^2}$$

$$= 0.999$$

b The acoustic impedance for gel is now $Z_1 = 1.65 \times 10^6\,\text{kg m}^{-2}\,\text{s}^{-1}$.
The acoustic impedance for skin is $Z_2 = 1.7 \times 10^6\,\text{kg m}^{-2}\,\text{s}^{-1}$.

$$\text{Fraction of ultrasound reflected} = \frac{(Z_2 - Z_1)^2}{(Z_2 + Z_1)^2}$$

$$= \frac{(1.7 \times 10^6 - 1.65 \times 10^6)^2}{(1.7 \times 10^6 + 1.65 \times 10^6)^2}$$

$$= 2.23 \times 10^{-4}$$

When gel is applied between the transducer and the skin, most of the incident ultrasound penetrates the skin. Without gel, most of the ultrasound is reflected and hence lost for imaging purposes.

Table 9.1 Acoustic impedances Z of body tissues

Medium	**Acoustic impedance** Z/kg m^{-2} s^{-1}
Air	4.29×10^2
Blood	1.59×10^6
Water	1.50×10^6
Brain	1.58×10^6
Soft tissue	1.63×10^6
Bone (dense)	7.78×10^6
Muscle	1.70×10^6
Skin	1.70×10^6

When ultrasound hits a boundary between media of acoustic impedances Z_1 and Z_2, and the value of Z_1 differs considerably from the value of Z_2, a large fraction of the ultrasound is reflected. When the value of Z_1 is similar to the value of Z_2, only a small fraction of the incident ultrasound is reflected.

In order to use data from ultrasound for the purpose of imaging within the body, it is necessary to minimise reflection at the air–skin interface. This is achieved through the use of gel between the transducer and the skin, which eliminates air. The difference between the impedance Z_1 of the gel and Z_2 of skin is small and so most of the ultrasound incident on this boundary penetrates into the body (see Worked Example 9.1).

Pulse echo measurement

The thickness of a medium may be measured using ultrasound reflections in much the same way that sonar echoes are used to measure the depth of an object beneath a ship (Figure 9.7). The total distance travelled by a sonar pulse is twice the depth of the object beneath the ship. Thus if the speed of sound in water is known and the time taken for the sound to travel to the object and back is measured, the total distance travelled by the sound may be calculated.

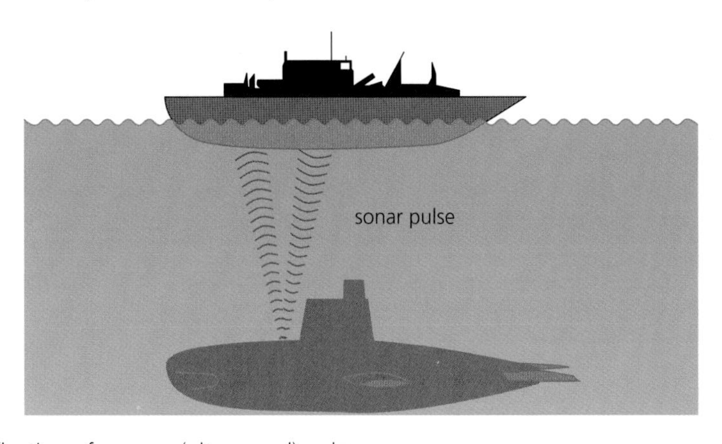

Figure 9.7 Reflection of a sonar (ultrasound) pulse

WORKED EXAMPLE 9.2

A ship sends a sonar pulse down to a submarine situated below it and detects the echo reflected from the submarine 0.085 s later. Calculate the depth of the submarine below the ship if the speed of sound in water is $1500\,\mathrm{m\,s^{-1}}$.

$$\begin{aligned}\text{Total distance travelled by sound} &= \text{speed of sound} \times \text{time of travel}\\ &= 1500\,\mathrm{m\,s^{-1}} \times 0.085\,\mathrm{s}\\ &= 127.5\,\mathrm{m}\end{aligned}$$

The sound has travelled down to the submarine and back. The depth of the submarine is therefore half of the total distance travelled by the sound:

$$\text{depth} = \frac{127.5\,\mathrm{m}}{2}$$

$$= 63.8\,\mathrm{m}$$

Ultrasound scans

The A-scan

When an ultrasound pulse is sent into a patient's body, signals due to reflections from the boundaries between different interfaces within the body are received at the transducer. The pulsed p.d. at the transducer is amplified and put across the Y-plates of a cathode ray oscilloscope (CRO). As the beam moves across the CRO screen, the pulses received are displayed against time. This type of display is called an **A-scan** (Figure 9.8).

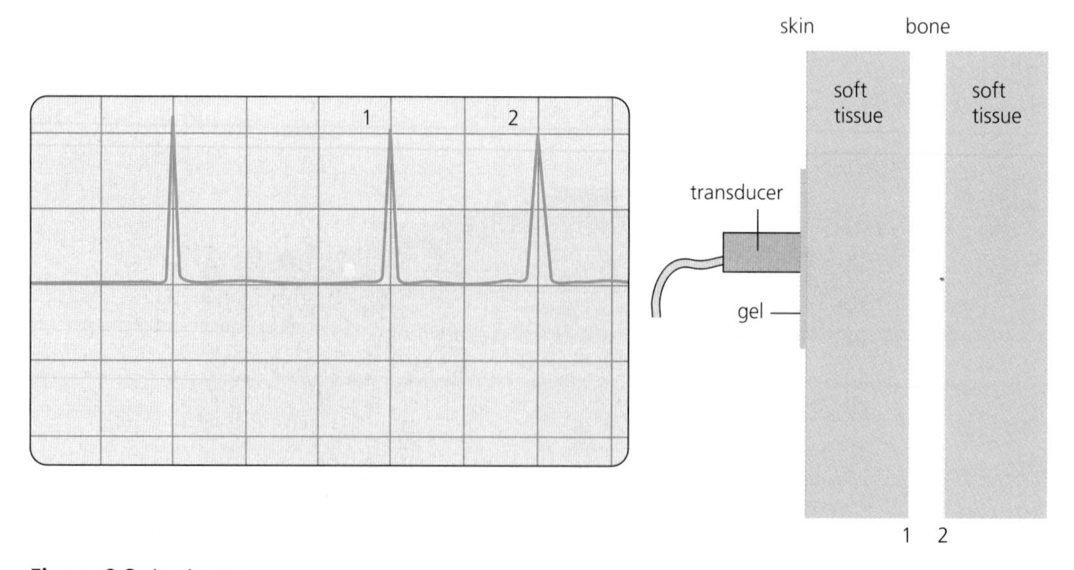

Figure 9.8 An A-scan

A-scans were the first type of ultrasound scans made. **B-scans** are based on similar principles and will be discussed later (page 136).

Measurements made from an A-scan can be used to determine the thickness of organs and bones. The time interval between two successive peaks on an A-scan trace is the time taken for the ultrasound to travel from the first boundary to the second boundary and back again. In Figure 9.8, peak 1 is due to reflection of an ultrasound pulse from the front edge of the bone while peak 2 is due to reflection of the pulse from the back edge of the bone. The distance travelled by the ultrasound pulse in the bone is then found by multiplying the speed of ultrasound in the bone by the time interval between the two peaks. The thickness of the bone is determined by dividing the total distance travelled by the ultrasound in the bone by 2.

WORKED EXAMPLE 9.3

The time-base setting of the CRO used for the A-scan in Figure 9.8 is $1.0 \times 10^{-5}\,\mathrm{s\,cm^{-1}}$. The distance between peaks 1 and 2 on the scan is 2 cm. Ultrasound travels at a speed of $4.0 \times 10^{3}\,\mathrm{m\,s^{-1}}$ in bone. What is the thickness of the bone?

The time interval between peaks 1 and 2 is

$$t = 2\,\mathrm{cm} \times 1.0 \times 10^{-5}\,\mathrm{s\,cm^{-1}} = 2.0 \times 10^{-5}\,\mathrm{s} \text{ or } 0.020\,\mathrm{ms}$$

The distance travelled by the ultrasound in the bone is given by:

$$\begin{aligned} s &= c \times t \\ &= 4.0 \times 10^{3}\,\mathrm{m\,s^{-1}} \times 2.0 \times 10^{-5}\,\mathrm{s} \\ &= 0.080\,\mathrm{m} \end{aligned}$$

The thickness of the bone is therefore

$$\frac{s}{2} = \frac{0.080\,\mathrm{m}}{2} = 0.040\,\mathrm{m}$$

A fraction of the incident ultrasound is reflected at each boundary and so the signal which penetrates further into the body is less intense. This means that subsequent reflections from deeper boundaries will produce weaker reflected signals at the transducer. Ultrasound is also attenuated in the medium through which it travels, reducing further the intensity of the final pulse received at the transducer. It is therefore necessary in thickness measurements to amplify the received signals prior to display on an A-scan. Signals from deep inside the body are amplified more than the signals from boundaries close to the surface. The relative magnitudes of the reflected signals from boundaries *prior* to amplification can give further information about the media through which the ultrasound is travelling.

A-scans are most commonly used today to enable the thickness of the eye lens to be measured prior to surgery.

The B-scan

The B-scan uses the same principle as the A-scan. The intensity of the reflected signal is represented not as a p.d. on the Y-plates of an oscilloscope (i.e. a peak in the trace) but as the brightness of a spot (Figure 9.9).

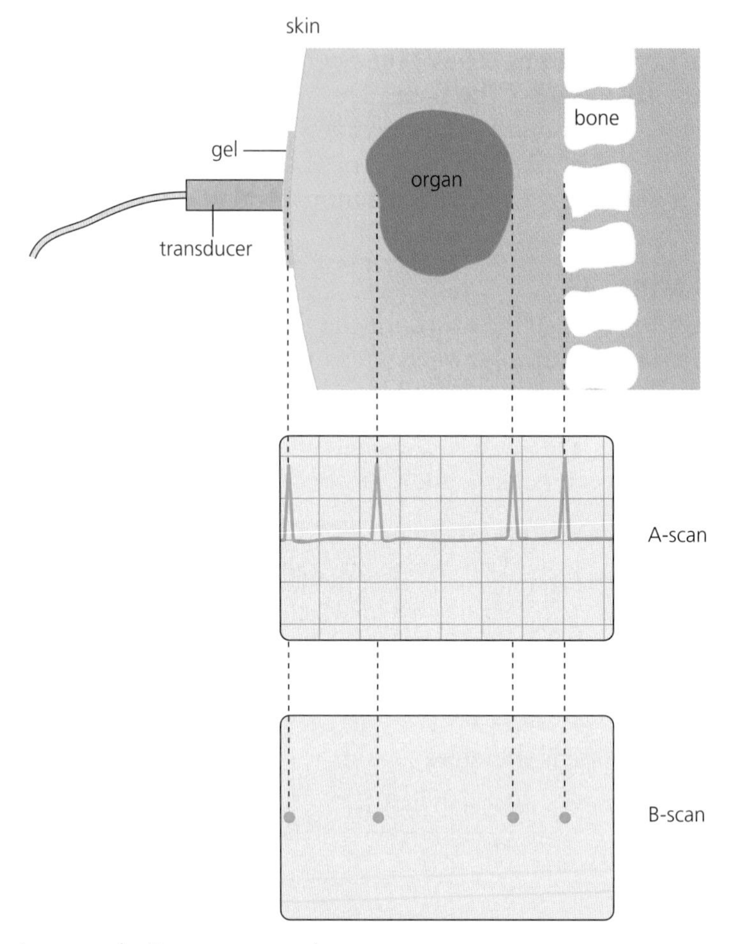

Figure 9.9 An A-scan and a B-scan compared

The ultrasound transducer is then re-angled so that the next pulse is incident along a slightly different direction. This produces a series of spots that are displaced from the first series and so the shape of the boundaries is defined. Figure 9.10 shows in outline the principle of the method. In practice, the resolution of the image is improved when the frequency of the pulses produced by the transducer is increased. The increased density of dots formed produces an image that resembles a low-quality photograph.

In some modern B-scanners the movement of the transducer is controlled by a computer, while in others there are a number of transducers each aligned in a linear array and switched on consecutively (Figure 9.11).

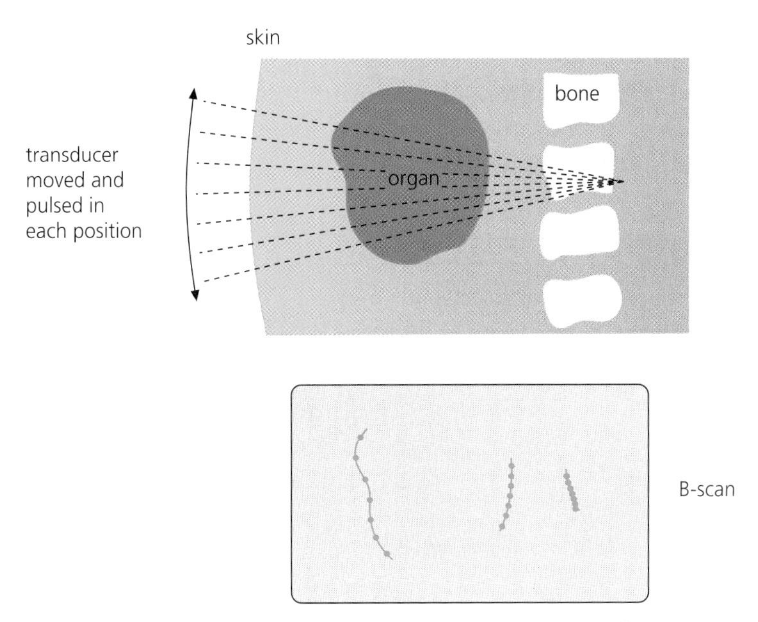

Figure 9.10 The principle of image production by a B-scan

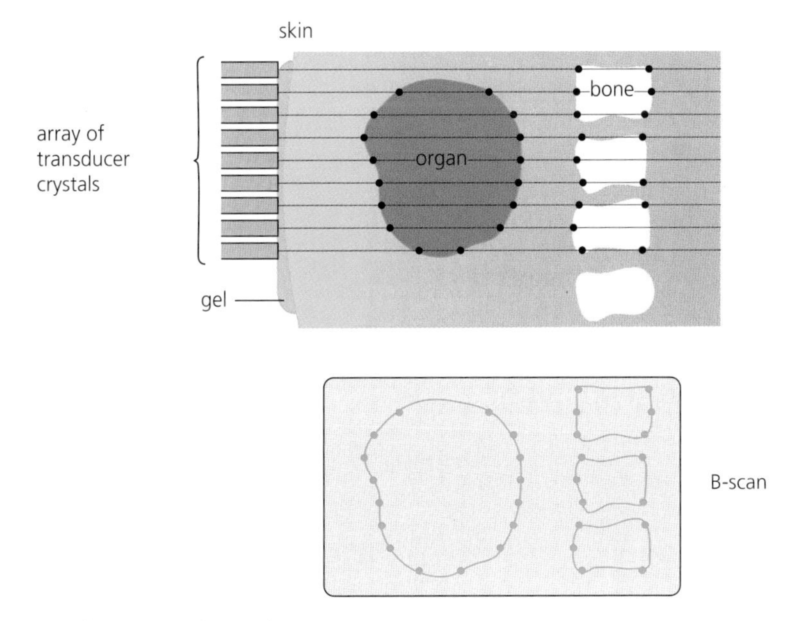

Figure 9.11 Use of an array of transducers

B-scans are most commonly used to monitor the developing foetus in the uterus (see Figure 9.1). Abnormalities in the development of the foetus may be diagnosed early, enabling planning for future treatment. Measurements are also made of the diameter of the head and length of the developing foetus, usually at about 16 weeks and 32 weeks into the pregnancy, to determine more accurately the number of weeks remaining till the birth.

Doppler ultrasound

Ultrasound can be used to measure blood flow rate. When ultrasound is pulsed and sent towards an artery or vein, the reflected pulse from the moving blood undergoes a change in frequency due to the **Doppler effect** (see Box 9.1).

Box 9.1 The Doppler effect

Consider a source that is generating a signal of frequency 10 Hz, first while stationary (Figure 9.12a) and then while moving (Figure 9.12b) towards a detector.

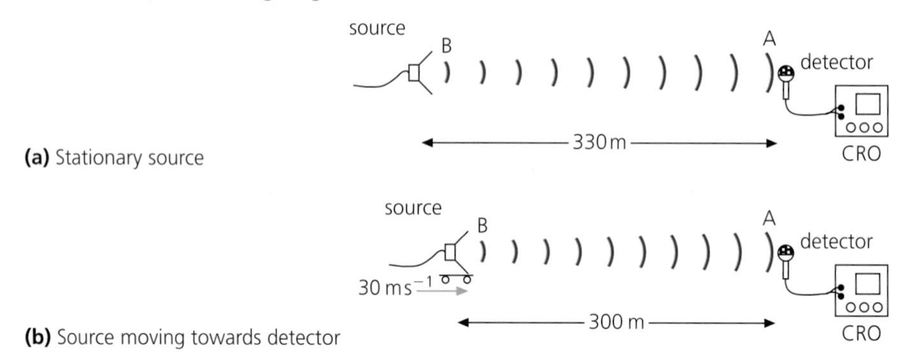

Figure 9.12

Figure 9.12a shows the source emitting ten compression waves in one second. These compressions travel from the source towards a detector positioned 330 m away. The tenth compression, B, is produced just as the first compression, A, reaches the detector. The wavelength of the waves may be calculated as:

$$\lambda = \frac{\text{distance from source to detector}}{10} = \frac{330\,\text{m}}{10} = 33\,\text{m}$$

If the speed of the signal in air is taken as the speed of sound, $c = 330\,\text{m}\,\text{s}^{-1}$, the frequency of the source may be verified using the equation:

$$f = \frac{c}{\lambda}$$

$$= \frac{330}{33} = 10\,\text{Hz}$$

Figure 9.12b shows the source emitting ten compressions in one second while moving towards the detector with a speed of $30\,\text{m}\,\text{s}^{-1}$. The first compression, A, reaches the detector as the tenth compression, B, leaves the source. At this moment in time the source is 300 m from the detector. This means that the ten compressions are now spread over a distance of 300 m. The wavelength λ of the waves measured by the detector is now

$$\lambda = \frac{300}{10} = 30\,\text{m}$$

The observed frequency f at the detector is

$$f = \frac{c}{\lambda}$$

$$= \frac{330}{30} = 11\,\text{Hz}$$

The pitch of a sound is observed to be higher when the sound source moves towards a stationary detector or, more simply, when the distance between source and detector shortens.

When the ultrasound transducer is stationary and the blood is moving towards the transducer, there is a shortening of the observed wavelength prior to reflection at the blood, due to the decreasing distance between the source and blood. There is also a second shortening of the wavelength because the reflector (the blood) acts as a moving source. This effect doubles the change in observed frequency. If the transducer is positioned so that the path of the emitted ultrasound makes an angle θ with the direction of the moving blood (Figure 9.13), the change in observed ultrasound frequency, Δf, is given by:

$$\Delta f = \frac{2fv\cos\theta}{c}$$

where f is the source (transducer) frequency, v is the speed of the moving blood, and c is the speed of ultrasound in the medium through which it passes.

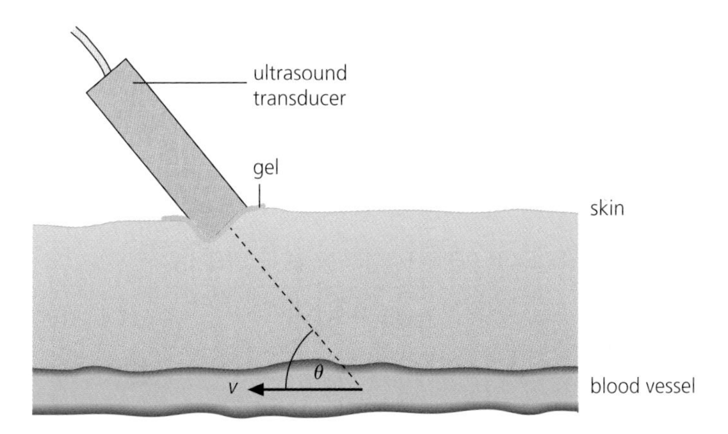

Figure 9.13

By measuring the difference Δf, between the reflected frequency and the source frequency, caused by the blood moving in veins or arteries towards the transducer, the average speed of the pulsating blood may be calculated (Figure 9.14). Doppler ultrasound techniques are also used to monitor foetal heartbeat.

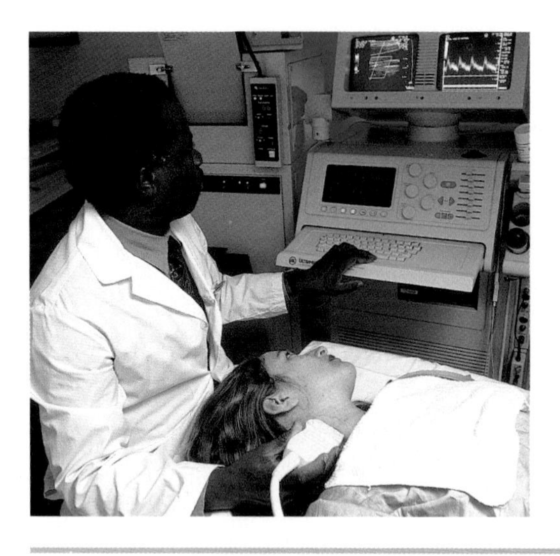

Figure 9.14 Measuring blood flow rate in a patient's artery in the neck by Doppler ultrasound scanning

WORKED EXAMPLE 9.4

The blood flow rate of a patient was measured using Doppler ultrasound. The frequency of the ultrasound used was 6.5 MHz and this was pulsed from a transducer so that the angle between the path of the ultrasound and the blood vessel was 60°. The resulting observed frequency of the reflected ultrasound pulse differed from the frequency of the incident pulse by 0.90 kHz. The speed of ultrasound in the tissue between the transducer and the blood vessel was $1.5\,\mathrm{km\,s^{-1}}$.

a Calculate the speed of blood flow in the blood vessel.

b Calculate the blood volume flow rate. The diameter of the blood vessel was 2.0 mm.

a Using the formula

$$\Delta f = \frac{2fv\cos\theta}{c}$$

and rearranging:

$$v = \frac{\Delta f \times c}{2f\cos\theta}$$

$$= \frac{0.90 \times 10^3\,\mathrm{s^{-1}} \times 1.5 \times 10^3\,\mathrm{m\,s^{-1}}}{2 \times 6.5 \times 10^6\,\mathrm{s^{-1}}} \times \cos 60°$$

$$= 0.21\,\mathrm{m\,s^{-1}}$$

b Volume flow rate = volume of blood that flows in 1 second
= cross-sectional area of vessel × speed of blood
$= \pi r^2 \times v$

$$= \pi\left(\frac{2.0 \times 10^{-3}}{2}\right)^2 \mathrm{m^2} \times 0.21\,\mathrm{m\,s^{-1}}$$

$$= 6.6 \times 10^{-7}\,\mathrm{m^3\,s^{-1}}$$

Questions

1 **a** Explain how a piezoelectric crystal is used to generate ultrasound.
b Describe the principle by which an image may be formed of a structure within the body using ultrasound.
c Explain the limitations of this process of imaging.

2 When an ultrasound pulse passes into a patient's body, it progresses through a soft tissue medium, into a bone and then back into a soft tissue medium.
a Using data from Table 9.1, calculate:
i) the fraction of the intensity of ultrasound that is reflected at the front edge of the bone,
ii) the fraction of the intensity of ultrasound that penetrates through the bone (assuming that the attenuation of ultrasound in the bone is negligible),
iii) the fraction of the ultrasound pulse transmitted through the bone that is reflected from the back edge of the bone,
iv) the fraction of the intensity of ultrasound (originally incident on the front edge of the bone) that eventually penetrates into the soft tissue after emerging from the back edge of the bone.
b State the process adopted to overcome the problems of the progressively smaller intensities of ultrasound that penetrate into the body at each boundary, when processing the reflected signals.

3 When an ultrasound pulse reflects from the front and back edges of an organ, it produces two peaks on an A-scan. The distance between these two peaks on the A-scan is measured as 3.4 cm on a CRO. The time-base for this CRO is set to $0.050\,\mathrm{ms\,cm^{-1}}$. If the speed of ultrasound in the organ is $1200\,\mathrm{m\,s^{-1}}$, calculate the thickness of the organ.

4 Ultrasound of frequency 5.0 MHz was pulsed from a transducer, inclined such that the angle between the incident ultrasound and an artery was 45°. The resulting observed frequency of the reflected ultrasound pulse differed from the frequency of the incident pulse by an average of 0.85 kHz. For this question take the speed of ultrasound in the tissue between the transducer and the artery as $1.5\,\mathrm{km\,s^{-1}}$.
a Calculate the speed of blood flow in the artery.
b Suggest why the reflected ultrasound pulses contain a range of frequencies.

5 The acoustic impedances for air, skin and gel are $430\,\mathrm{kg\,m^{-2}\,s^{-1}}$, $1.7 \times 10^6\,\mathrm{kg\,m^{-2}\,s^{-1}}$ and $1.6 \times 10^6\,\mathrm{kg\,m^{-2}\,s^{-1}}$ respectively.
a Calculate the fraction of ultrasound that penetrates an air–skin boundary.
b Calculate the fraction of ultrasound that penetrates a gel–skin boundary.
c Use your answers to **a** and **b** to explain why gel is put onto the skin prior to using an ultrasound transducer.

Lasers and endoscopes

In this chapter you will read about:

- the structure of a laser
- the uses of lasers in medicine
- the principle of total internal reflection applied to glass fibres
- the structure and workings of an endoscope

Figure 10.1 The intense beam produced by a laser

The laser

A laser produces an intense beam of coherent, monochromatic light. A summary explanation of the production of such a beam may be found in the words from which the acronym **laser** is derived: **l**ight **a**mplification by **s**timulated **e**mission of **r**adiation.

To understand in more detail the processes involved in the production of laser light, it is first necessary to revisit the mechanism whereby photons are absorbed by atoms. When a photon of electromagnetic radiation is incident on an atom, the atom will absorb the photon only if the energy of the photon corresponds to a jump by the atom to one of its allowed energy levels. These energy levels are specific to atoms of a given material. If absorption occurs, the absorbing atom is said to be excited and it remains in this state for a period of less than 1 μs. As the atom returns to its unexcited (or ground) state, it emits a photon of radiation equal in energy to the difference in energy levels of the atom before and after the jump.

When atoms in a material are exposed to photons of electromagnetic radiation, they are constantly absorbing and emitting photons. Equilibrium is reached within the material when the number of photons absorbed every second equals the number of photons emitted every second. At any given moment in time, most of the atoms will be found in the ground state. When an excited atom jumps down to the ground state, it may do so in several jumps, emitting photons of different energies (Figure 10.2a–c).

For atoms of certain materials, there exists an excited state known as the **metastable** state. This is the state in which the atom is found just before it finally decays to the ground state (Figure 10.3).

An atom remains in the metastable state longer than it does in any other of its excited states. This means that at any one moment in time there will be a greater number of atoms in the metastable state than in any other of the excited states. Atoms in the metastable state are **stimulated** into decaying back to the ground state by interaction with a photon of energy $E_1 - E_0$ (the difference between the energy in the metastable state, E_1, and the energy in the ground state, E_0). This produces a photon of this same energy, which in turn goes on to stimulate other atoms that are in the metastable state, resulting in a cascade effect.

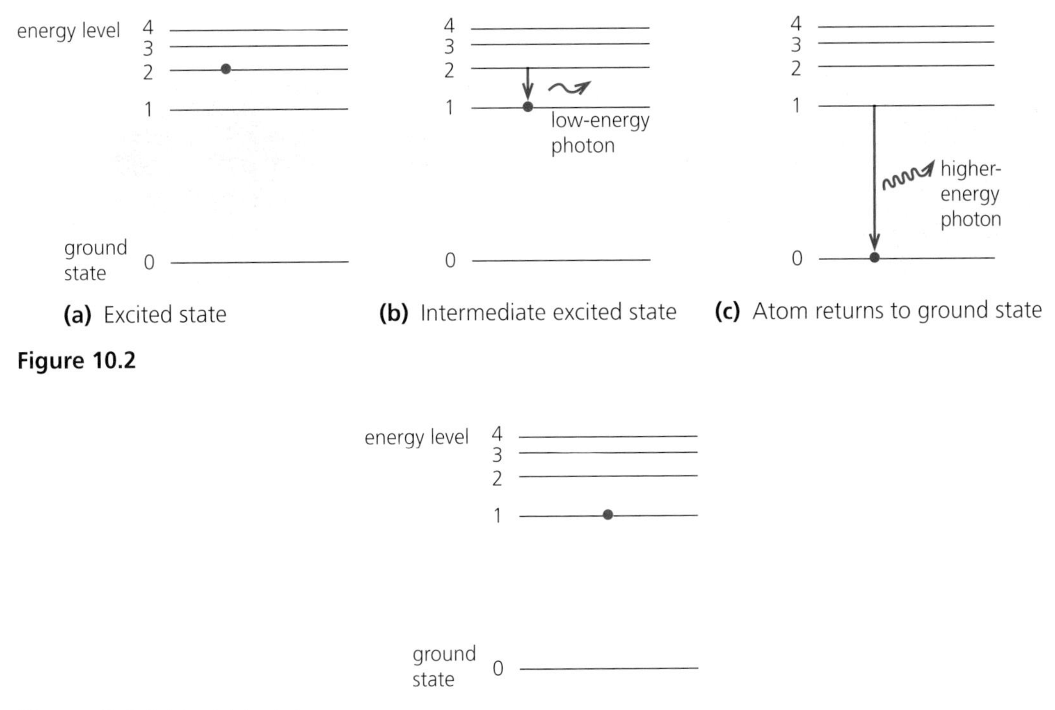

Figure 10.2

Figure 10.3 Metastable state

A laser uses a material which allows this effect. At the ends of the laser tube are mirrors. These reflect photons back into the atoms of the emitting medium, so stimulating the production of yet more photons (Figure 10.4). All of these stimulated photons are coherent (that is, they are all in phase). The end result is a beam of very intense, coherent, monochromatic light.

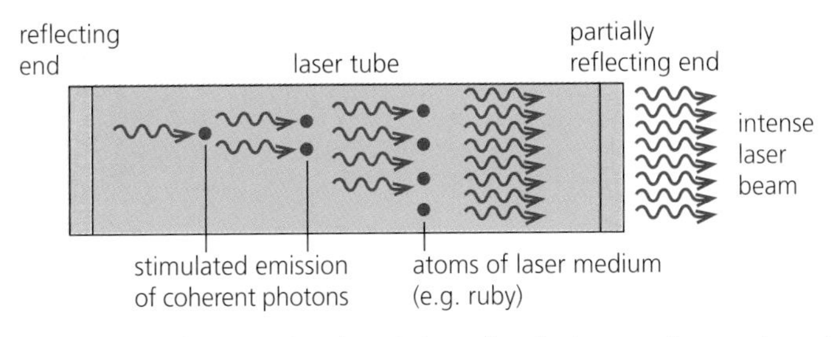

Figure 10.4 Light amplification by stimulated emission of radiation produces a laser beam

The uses of lasers in medicine

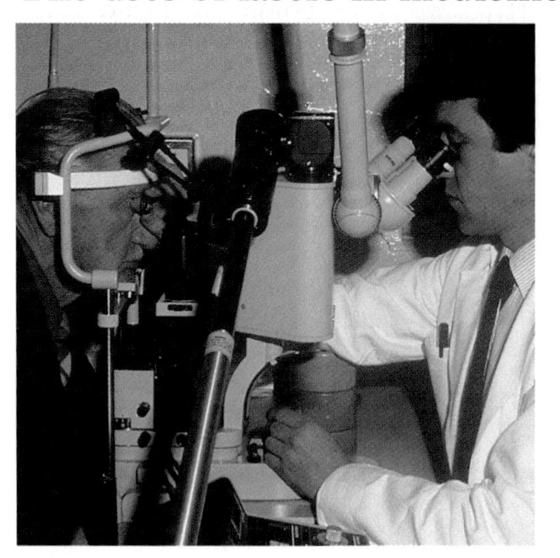

Figure 10.5 Patient undergoing laser treatment of the cornea

Lasers are employed today in a variety of ways in medicine. In certain surgery, the laser has completely replaced the conventional scalpel. It is also used to re-attach detached retinas, to reshape corneas (Figure 10.5), to remove birthmarks, and in therapy to treat certain tumours (see below).

As a scalpel

Lasers may be used in preference to conventional scalpels to cut tissue. The advantages of a laser in this application are that the incision is sterile, the cut is very fine and the depth of the incision accurately controllable. The energy from the laser radiation vaporises the water content of cells in its path, causing the cells to shrink and thus cauterising blood vessels. This reduces the amount of bleeding during surgery.

In therapy

Laser light is absorbed to a greater degree by malignant tissue than by healthy tissue. The consequence is that during treatment of a tumour, surrounding healthy tissue suffers less damage. The colour of the tumour is important in the consideration of the wavelength of the laser used in the treatment. A tumour of a dark pigment will absorb the laser radiation more readily than a white tumour. The energy absorbed per kilogram of tissue and the time for which the laser energy is delivered are carefully calculated before treatment is applied. The intensity of laser light at a given point is given by the equation:

$$\text{intensity} = \frac{\text{power}}{\text{area}}$$

If the laser light from a parallel beam of given cross-sectional area is focused onto a surface that has a smaller area, the intensity of the laser light at the surface is increased. This means that the rate of delivery of energy per unit area at the surface is increased. Similarly, if the same laser light is focused onto a large area, the energy is distributed over this large area so the energy per unit area is decreased.

When treating superficial tumours, the laser beam is focused onto a small area using a convex lens of high power. The high power of the lens causes the intensity of the beam to decrease by a large factor over a short distance either side of the focal point and this reduces damage to the surrounding healthy tissue. For example, Figure 10.6a shows a 10 W laser beam with an objective lens of low power, and Figure 10.6b shows the same laser with an objective lens of high power. The intensity of the light 2 mm below the focal point is much greater in (a) than the intensity 2 mm below the focal point in (b).

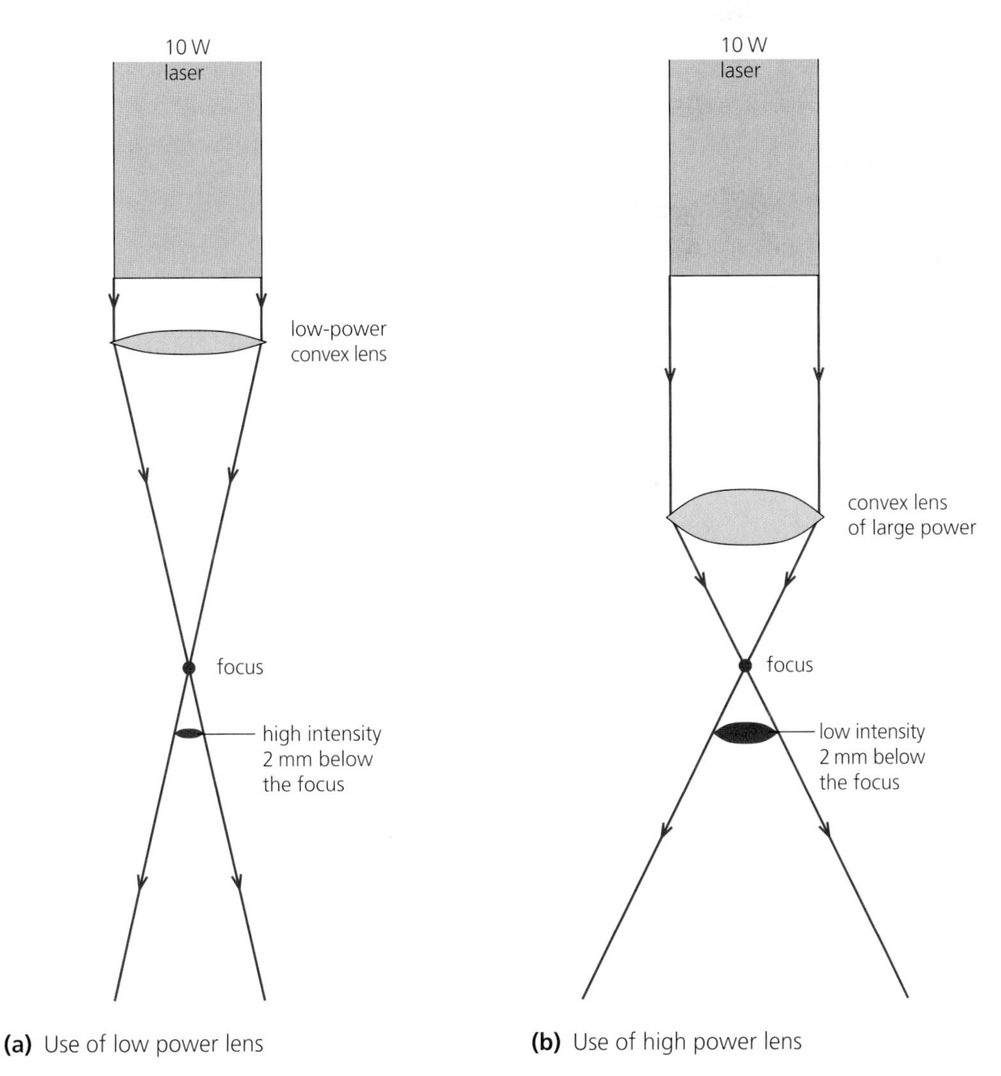

(a) Use of low power lens

(b) Use of high power lens

Figure 10.6

WORKED EXAMPLE 10.1

During laser treatment to re-attach a detached retina, a ruby laser is used. The energy of each laser pulse absorbed by an area of the retina of $1.2 \times 10^{-3}\,\text{mm}^2$ is 0.085 J. The duration of each pulse is 0.80 ms.

a Calculate:
  i) the minimum power of the laser,
  ii) the intensity at the retina.

b The threshold time for the sensation of pain is about 180 ms of laser treatment at this intensity. Calculate the energy delivered by the laser at the retina if pain is just experienced.

a i) Power = rate of conversion of energy $= \dfrac{\text{energy converted}}{\text{time taken}}$

$$= \frac{0.085\,\text{J}}{0.80 \times 10^{-3}\,\text{s}}$$

$$= 106\,\text{W}$$

ii) Intensity $= \dfrac{\text{power}}{\text{area}}$

$$= \frac{106\,\text{W}}{1.2 \times 10^{-9}\,\text{m}^2} \qquad (1\,\text{mm}^2 = 1 \times 10^{-6}\,\text{m}^2)$$

$$= 8.9 \times 10^{10}\,\text{W}\,\text{m}^{-2}$$

b Power = 106 W, and the time for which the laser is pulsed = 180 ms.
Energy absorbed = power × time

$$= 106\,\text{W} \times 180 \times 10^{-3}\,\text{s}$$

$$= 19\,\text{J}$$

Transmission of light through a glass fibre

Figure 10.7 shows an incident ray of light entering a straight section of glass fibre. As the angle of incidence, *i*, for the ray inside the fibre as it strikes the edge of the fibre is greater than the critical angle, *c*, for the glass fibre, the ray undergoes **total internal reflection** (see Box 10.1).

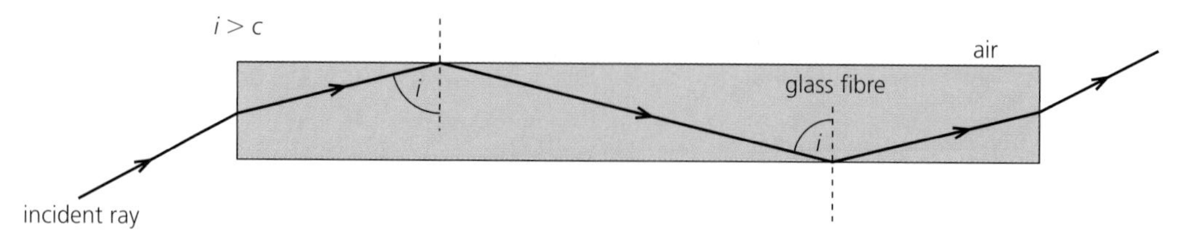

Figure 10.7 Total internal reflection in a glass fibre

Box 10.1 Total internal reflection

When light travels from one transparent medium into another transparent medium made from a different material, refraction occurs at the boundary. Light is deviated from its original path. The degree of deviation is determined by the angle of incidence, *i*, and a property of both media on either side of the boundary called the **refractive index**.

Figure 10.8 shows a ray of light travelling from a medium of refractive index n_1 into a medium of refractive index n_2. The angle between the incident ray and the normal to the boundary is the angle of incidence, *i*. The angle between the refracted ray and the normal is the angle of refraction, *r*.

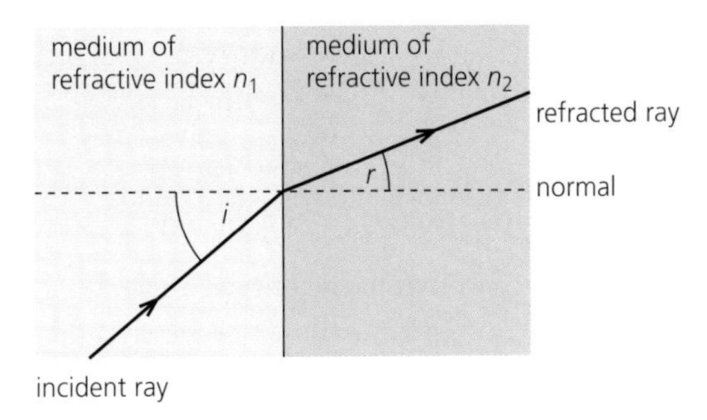

Figure 10.8 Refraction of light at a boundary between different media

The angles *i* and *r*, and the refractive indices n_1 and n_2 of the media either side of the boundary are related by the equation:

$$n_1 \sin i = n_2 \sin r$$

When light passes from an optically less dense medium (such as air) into an optically denser medium (such as glass) it is refracted towards the normal. Conversely, when light passes from an optically denser medium (such as glass) into an optically less dense medium (such as air) it is refracted away from the normal (Figure 10.9).

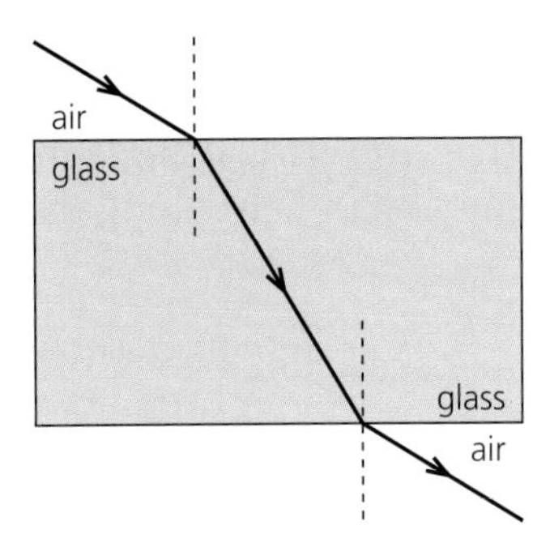

Figure 10.9 Refraction at air–glass and glass–air boundaries

When the angle of incidence for light passing from a denser medium into a less dense medium increases, so the angle of refraction increases (Figures 10.10a and b). The value of the incident angle that results in an angle of refraction of value 90° is known as the **critical angle**, *c*, for a given boundary and a given colour of light. For light incident on a boundary at the critical angle, refraction occurs in a direction parallel to the boundary between the two media (Figure 10.10c).

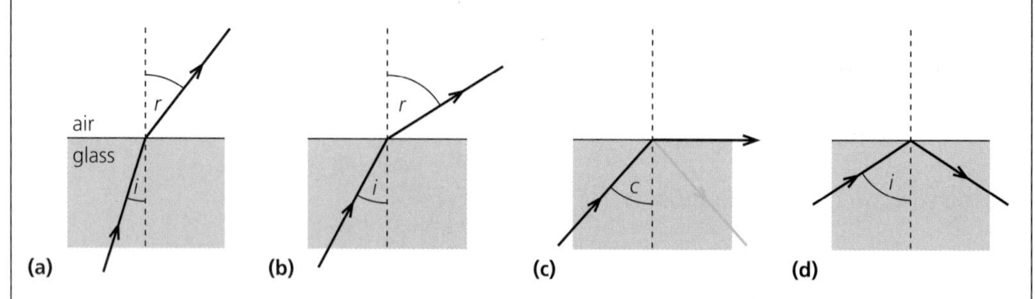

Figure 10.10 Increasing the angle of incidence at a glass–air boundary

For light incident at angles greater than that of the critical angle, total internal reflection takes place (Figure 10.10d). For these angles, no refraction occurs and so no light escapes from the denser medium.

Light can pass through a very long glass fibre in air by total internal reflection all the way along the fibre. This process occurs as long as the incident beam makes an angle of incidence greater than that of the critical angle for the glass–air boundary. Any grease or dirt on the outside surface of the glass fibre will change the refractive index of the glass fibre (and hence the critical angle) and result in the loss of light from the fibre through refraction. Similarly, scratches on the outside surface of the fibre will reduce the angle of incidence of the internal ray and so less light will be transmitted through the fibre.

Glass fibres used in fibre optic cables are usually only a few μm in diameter and are coated in a protective layer made of a material of refractive index lower than that of the glass. When the angle of incidence at the fibre–coating boundary is equal to the critical angle *c* for this interface (Figure 10.11), some light escapes from the fibre. For larger angles of incidence *i* at the air–fibre boundary, the angle of incidence at the fibre–coating boundary is less than the critical angle, light is increasingly refracted through the coating and transmission of light along the fibre ceases.

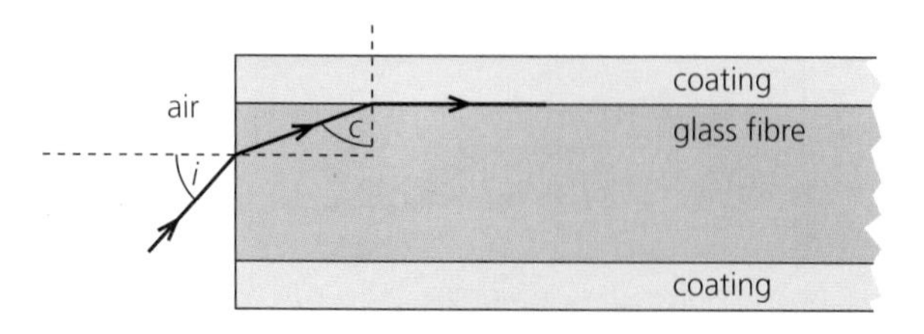

Figure 10.11 The critical angle for a coated fibre

WORKED EXAMPLE 10.2

Light enters a straight section of glass fibre of refractive index 1.50 from air of refractive index 1.00, and total internal reflection takes place within the fibre. The refractive index of the coating around the fibre is 1.42. Calculate:

a the critical angle for the *fibre–coating* boundary,
b the angle of refraction at the *air–fibre* boundary that causes light to be incident at the critical angle at the fibre–coating boundary,
c the maximum angle of incidence at the *air–fibre* boundary that just causes light to escape the fibre and travel parallel to the fibre (as in Figure 10.11).

a At the fibre–coating boundary:

$$n_f \sin i = n_c \sin r$$

$n_f = 1.50$, $n_c = 1.42$, and $r = 90°$ when $i = c$.

So

$$\sin c = \frac{1.42 \sin 90°}{1.50} = 0.9466$$

and

$$c = 71.2°$$

b At the air–fibre boundary r is given by (see Figure 10.11):

$$r = 90° - c = 90° - 71.2° = 18.8°$$

c At the air–fibre boundary:

$$n_a \sin i = n_f \sin r$$

$$\sin i = \frac{n_f \sin r}{n_a}$$

When $r = 18.8°$:

$$\sin i = \frac{1.5 \sin 18.8°}{1.0} = 0.48$$

and so

$$i = 28.9°$$

For angles of incidence at the air–fibre boundary greater than this, there is no total internal reflection within the fibre and transmission along the fibre stops.

The endoscope

Endoscopes are commonly used to directly view internal parts of the body and to perform minor operations (Figure 10.12). They basically consist of long glass fibres arranged in one of two ways, forming either an **incoherent** bundle (see Figure 10.13a opposite) or a **coherent** bundle (Figure 10.13b).

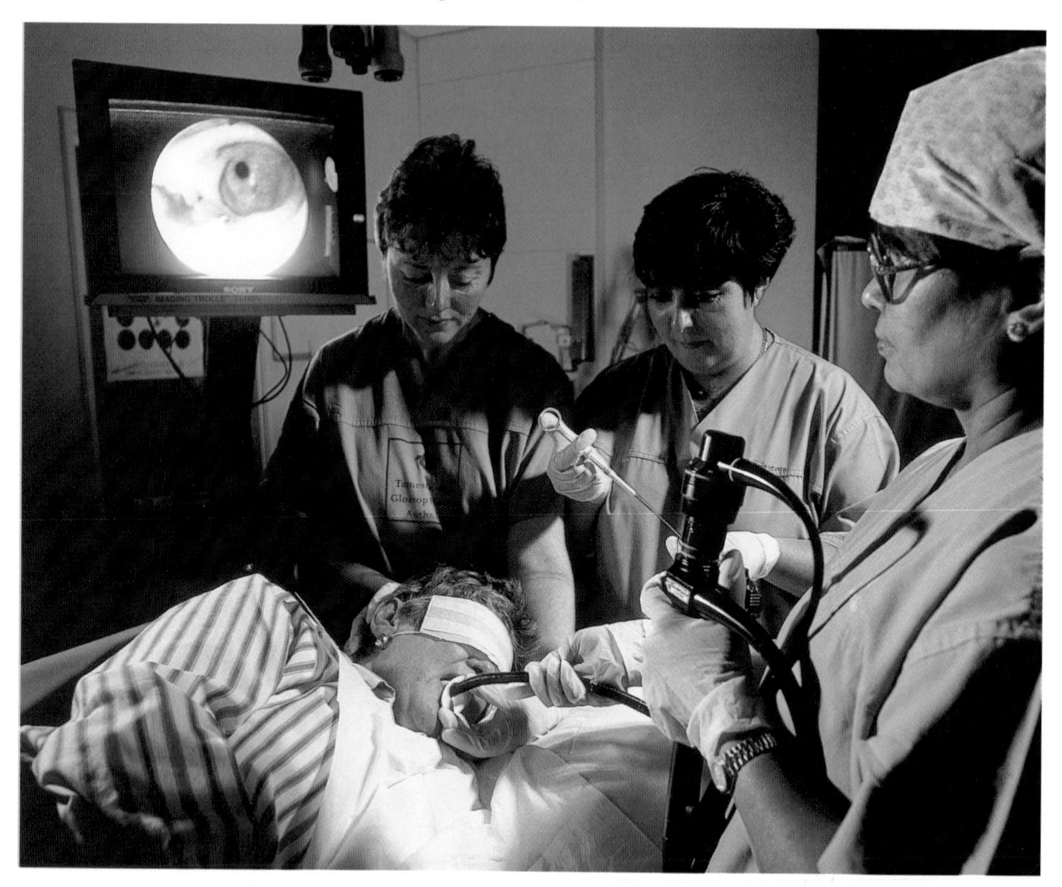

Figure 10.12 An endoscope in use to view a patient's stomach and take a sample of tissue. The image obtained is seen on the screen

The arrangement of fibres at one end of a coherent bundle of fibres exactly matches the arrangement of the fibres at the other end. It is more expensive to produce a coherent bundle of fibres. The arrangement of fibres at one end of an incoherent bundle of fibres differs from that at the other end because the strands of fibres are tangled when put together.

In order to illuminate an object within the body, it is necessary to carry light to the object. This is achieved by passing light along an incoherent bundle of fibres. Light emerges from the fibres and reflects off the object being viewed. The reflected light passes through an objective lens, returns along a coherent bundle of fibres and is focused by an eye-piece lens to form an image.

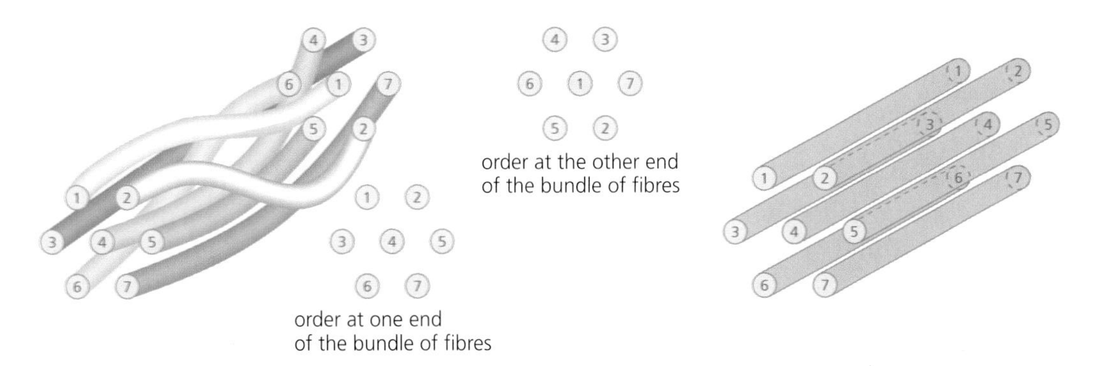

(a) Incoherent bundle of fibres

(b) Coherent bundle of fibres

Figure 10.13

A typical endoscope consists of a flexible cylindrical tube (Figure 10.14) with an approximate length and diameter of 1.5 m and 10 mm respectively. At the operator's end of the tube are a series of controls which operate the movement of the other end of the endoscope (which is called the **distal tip**) and allow the cutting and collection of samples of tissue.

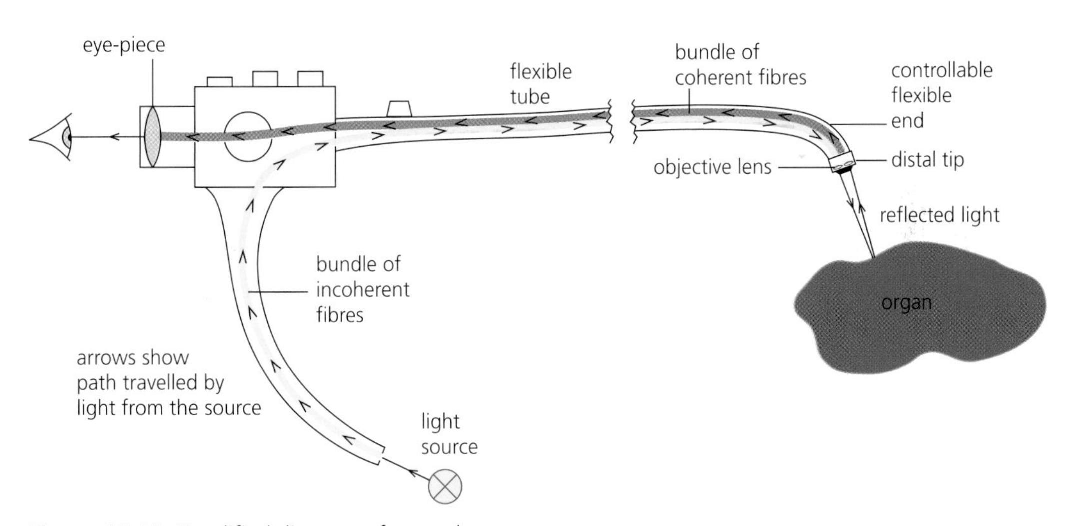

Figure 10.14 Simplified diagram of an endoscope

The cylindrical tube carries bundles of coherent and incoherent fibres, a narrow tube along which water may be pumped to wash the objective lens, a channel for the removal of fluids from the body and various channels to operate the tools that might be used.

Light is attenuated in glass fibres. Over a length of several metres of glass fibre, up to 50% of the light incident at one end of the fibre may be attenuated. This means that the light source at one end of the incoherent bundle of fibres in an endoscope must be very intense if light is to emerge from the other end.

The glass fibres in the coherent bundle that carry light from inside the patient to the viewer are smaller in diameter than those in the incoherent bundle (typically of the order of 10 μm). The greater the number of these fibres per cross-sectional area of the bundle, the greater the resolution of the final image formed by the endoscope. If the diameter of the fibres is reduced to values below 1 μm, however, diffraction of light occurs at the end of the fibres and some of the incident light escapes through refraction.

Figure 10.15 An endoscope image showing the removal of a kidney stone by jaws at the end of a second endoscope. The jaws are operated by the surgeon remotely using scissor handles

A major strength of the endoscope is that it allows many exploratory procedures to be carried out without the need for surgery. This reduces the risk to a patient and decreases the recovery period. Where there is a natural opening in the body, the endoscope may be simply inserted, permitting for example the oesophagus, stomach or colon to be viewed. In the absence of a natural opening near to the organ that is to be investigated, a small incision may be made allowing entry for the endoscope. In this way it is possible to inspect organs such as the liver. With the addition of accessories such as scalpels and forceps, minor operations such as biopsies (the removal of small slices of tissue for analysis) and the removal of an obstruction such as a kidney stone (Figure 10.15) may be carried out.

Questions

1 a State the meaning of the acronym LASER.

b Explain the need for a medium whose atoms have a metastable state if laser light is to be generated.

c Explain two ways in which the laser light is made more intense.

2 a Explain the principle by which light passes along a glass fibre.

b Describe the difference between a coherent and an incoherent bundle of fibres.

c Briefly describe the operation of an endoscope for the purpose of examining the gastrointestinal tract.

d State and explain a situation in which:

i) an examination using an endoscope might be employed instead of an X-ray,

ii) an X-ray might be employed instead of a visual examination using an endoscope.

3 a Describe the properties of a laser that makes it suitable for use as a scalpel.

b Explain the advantages of the use of a laser instead of a conventional scalpel to cut tissue.

c Explain why an objective lens of large power is used during the laser treatment of a superficial tumour.

4 A laser of power 50 W is focused onto a circular area of diameter 0.50 mm. The energy required at the site of the treatment is calculated as 0.15 J. Calculate:

a the intensity of the laser light at the circular area,

b the duration for which the laser should be pulsed to deliver 0.15 J of energy to the circular area.

5 A glass fibre in the core of a fibre optic strand has a refractive index of 1.52. The coating around the core has a refractive index of 1.40. Air has a refractive index of 1.0. The glass fibre lies on a horizontal surface and is straight.

a Calculate the critical angle for the boundary between the glass fibre and the coating.

b Calculate the critical angle for a boundary between glass fibre and air.

c Explain the reason for the coating that surrounds the fibre optic strand.

Answers

Chapter 1

1 a See Figure 1.8.

2 b i) 400–700 nm

3 a As power $= \dfrac{1}{\text{focal length } f}$ and

$$\frac{1}{\text{focal length } f} = \frac{1}{\text{object distance } u} + \frac{1}{\text{image distance } v}$$

so power $= \dfrac{1}{u} + \dfrac{1}{v}$

The object distance $u = 0.50$ m, and the image distance is the distance of the retina from the cornea, so $v = 0.019$ m.

$$\therefore \text{power} = \frac{1}{0.50} + \frac{1}{0.019}$$

$$= 54.6\,\text{D}$$

b When the eye focuses on an object at infinity, $u = \infty$

The power of the refracting system $= \dfrac{1}{u} + \dfrac{1}{v}$

$$= \frac{1}{\infty} + \frac{1}{0.019}$$

$$= 52.6\,\text{D}$$

c A lens of power x D must be added to the eye of power 54.6 D to bring the combined power to the required value of 52.6 D.
$54.6 + x = 52.6$
$\therefore x = -2.0\,\text{D}$

d Short sight (myopia); concave lens.

e Power $= \dfrac{1}{u} + \dfrac{1}{v}$

$$= \frac{1}{0.15} + \frac{1}{0.019}$$

$$= 59.3\,\text{D}$$

f The combined value of the power of the lens in **c** and the eye when viewing an object at the near point is $59.3\,\text{D} + (-2.0\,\text{D}) = 57.3\,\text{D}$.

Power $= \dfrac{1}{u} + \dfrac{1}{v}$

$$57.3 = \frac{1}{u} + \frac{1}{0.019}$$

$\therefore u = 0.21\,\text{m} = 21\,\text{cm}$

5 a Power $= \dfrac{1}{u} + \dfrac{1}{v}$ (see answer to **3a**)

$$= \frac{1}{0.25} + \frac{1}{0.018}$$

$$= 59.6\,\text{D}$$

b Power $= \dfrac{1}{u} + \dfrac{1}{v}$

$$= \frac{1}{0.60} + \frac{1}{0.018}$$

$$= 57.2\,\text{D}$$

c A lens of power x must be added to 57.2 D to bring the power to the required value of 59.6 D.
$x + 57.2 = 59.6$
$\therefore x = +2.4\,\text{D}$ (a convex lens)

Chapter 2

2 b i) About 25 Hz to about 16 kHz, depending on age.
ii) The hairs in the cochlea do not respond to lower frequencies; the ossicles do not respond to higher frequencies.
iii) In childhood, the range is about 25 Hz–16 kHz. Ageing causes the lower limit to rise and the upper limit to fall so that the range becomes about 30 Hz–8 kHz or even lower.

3 a i) See Figure 2.5.
ii) The line should always be above that in Figure 2.5, a slightly flatter curve starting at about 30 Hz and finishing at about 8 kHz.

b Loss $= 10\,\lg\dfrac{I}{I_0}$ where $I_0 = 1.0 \times 10^{-12}\,\text{W m}^{-2}$

$$= 10\,\lg\frac{3.2 \times 10^{-10}}{1.0 \times 10^{-12}}$$

$$= 25\,\text{dB}$$

4 a The ear's response to sound intensity is loudness.
For a sound to be twice as loud, the intensity is doubled.
For a sound to be three times louder, the intensity is quadrupled.
For a sound to be n times louder, the intensity is 2^{n-1} greater.
$\lg 2^n = n \lg 2$, so a logarithmic scale is used.

b i) The sound power is emitted uniformly over a sphere of radius 1.5 m.
Surface area of sphere $= 4\pi \times 1.5^2 = 28.3\,\text{m}^2$

$$\text{Intensity} = \frac{\text{power}}{\text{area}} = \frac{0.90}{28.3}$$

$$= 0.032\,\text{W m}^{-2}$$

ii) Intensity level $= 10\lg\frac{I}{I_0}$

where $I_0 = 1.0 \times 10^{-12}\,\mathrm{W\,m^{-2}}$, so

$$\text{intensity level} = 10\lg\frac{3.2\times10^{-2}}{1.0\times10^{-12}} = 105\,\mathrm{dB}$$

5 a Intensity level $= 10\lg\frac{I}{I_0}$

$$= 10\lg\frac{7.4\times10^{-3}}{1.0\times10^{-12}} = 98.7\,\mathrm{dB}$$

b Intensity level $= 10\lg\frac{I}{I_0}$

$$95 = 10\lg\frac{I}{1.0\times10^{-12}}$$

giving $I = 3.2\times10^{-3}\,\mathrm{W\,m^{-2}}$

c Intensity at 68 dB $= 6.3\times10^{-6}\,\mathrm{W\,m^{-2}}$
Intensity at 92 dB $= 1.6\times10^{-3}\,\mathrm{W\,m^{-2}}$

$$\text{Fractional change} = \frac{1.6\times10^{-3} - 6.3\times10^{-6}}{6.3\times10^{-6}} = 250$$

6 a ii) See Figure 2.7 (maximum at about 2 kHz, zero outside range 25 Hz–16 kHz).

b Intensity level $= 10\lg\frac{I}{I_0}$

where $I_0 = 1.0\times10^{-12}\,\mathrm{W\,m^{-2}}$

For 85 dB noise level, $85 = 10\lg\frac{I}{1.0\times10^{-12}}$

giving $I = 3.16\times10^{-4}\,\mathrm{W\,m^{-2}}$

For 87.5 dB noise level, $87.5 = 10\lg\frac{I}{1.0\times10^{-12}}$

giving $I = 5.62\times10^{-4}\,\mathrm{W\,m^{-2}}$

$$\text{Fractional change in intensity} = \frac{5.62 - 3.16}{3.16} = 0.78$$

Chapter 3

1 The distance from the base of the spine to the foot is about 90 cm = 0.90 m.

$$\text{Time taken} = \frac{\text{distance}}{\text{speed}} = \frac{0.9\,\mathrm{m}}{150\,\mathrm{m\,s^{-1}}} = 0.006\,\mathrm{s} = 6\,\mathrm{ms}$$

3 a See Figure 3.5.

b 10 pulses in $\frac{35}{5}\,\mathrm{s} = 7\,\mathrm{s}$

$\therefore$ pulse rate is $\frac{10}{7}\,\mathrm{s^{-1}} = \frac{600}{7}$ or 86 per minute.

4 a A: depolarisation of the auricles, causing them to contract.
B: depolarisation and contraction of the ventricles.
C: repolarisation and relaxation of the ventricles.

b i) Pulse heights reduced.
ii) Part of trace missing.

5 a *Systolic*: maximum blood pressure.
Diastolic: lowest pressure at which turbulence is heard in the artery.

b i) 165 is the systolic blood pressure measured in mm of mercury;
100 is the diastolic blood pressure measured in mm of mercury.
ii) Pressure = height × density × gravitational field strength

$$= 165\times10^{-3}\,\mathrm{m}\times13\,600\,\mathrm{kg\,m^{-3}}\times9.8\,\mathrm{N\,kg^{-1}} = 2.2\times10^{4}\,\mathrm{Pa}$$

Chapter 4

1 a For the arm in Figure 4.14 to be in equilibrium, the sum of the anticlockwise moments about the elbow joint J must equal the sum of the clockwise moments.
Clockwise moments about J $= B\times0.030\,\mathrm{m}$
Anticlockwise moments about J

$$= (15\,\mathrm{N}\times0.35\,\mathrm{m}) + (25\,\mathrm{N}\times0.25\,\mathrm{m})$$

$B\times0.030\,\mathrm{m} = 5.25\,\mathrm{N\,m} + 6.25\,\mathrm{N\,m}$

$\therefore B = 383\,\mathrm{N}$

b Ignoring the weight of the lower arm, the load is 15 N. The effort is 383 N.

$$\text{Mechanical advantage} = \frac{\text{load}}{\text{effort}} = \frac{15}{383} = 0.039$$

2 A

4 b A metabolic rate of 80 W means that energy is converted within the body for the purposes of maintaining metabolic processes at a rate of $80\,\mathrm{J\,s^{-1}}$.
8 hours $= 8\times60\times60\,\mathrm{s}$
Energy converted in 8 hours for maintaining metabolic processes $= 80\times8\times60\times60\,\mathrm{J} = 2.3\times10^{6}\,\mathrm{J}$
This is the minimum energy required when sleeping for this period.

5 For the head in Figure 4.16 to be in equilibrium, the clockwise and anticlockwise moments about the atlas vertebra must be equal.
effort $\times 0.040\,\mathrm{m} = 150\,\mathrm{N}\times0.025\,\mathrm{m}$
$\therefore$ effort = 94 N

Chapter 5

2 a μ is the linear absorption coefficient.

b HVT is that thickness of material required to reduce the intensity of the beam to one half of its initial value. It may be denoted by $x_{\frac{1}{2}}$.
Since $I = I_0 e^{-\mu x}$ and $I = \frac{1}{2}I_0$ when $x = x_{\frac{1}{2}}$, then
$\frac{1}{2} = \exp(-\mu x_{\frac{1}{2}})$
Inverting and taking logs (to base e),
$\ln 2 = \mu x_{\frac{1}{2}}$
$\therefore x_{\frac{1}{2}} = \dfrac{0.693}{\mu}$

c The attenuation coefficient then depends only on the nature and energy of the radiation, and not on the individual absorbing material.

3 b See Table 5.1.

c See Figure 5.5.

d Bone is more dense than muscle and contains many atoms with higher proton numbers. The combined attenuation coefficient at this photon energy is greater for bone than for muscle.

4 b 0.15 C of charge is
$\dfrac{0.15}{1.6 \times 10^{-19}} = 9.4 \times 10^{17}$ ion pairs

Energy required $= 9.4 \times 10^{17} \times 34\,\text{eV}$
$= 3.2 \times 10^{19}\,\text{eV}$
$= 3.2 \times 10^{19} \times 1.6 \times 10^{-19}\,\text{J} = 5.1\,\text{J}$
This is the energy deposited per kg of air.
Energy deposited in 2.5 g of air $= 5.1 \times 2.5 \times 10^{-3}\,\text{J}$
$= 1.3 \times 10^{-2}\,\text{J}$

Chapter 6

1 c Energy given to an electron
= charge on electron × accelerating voltage
$= 1.6 \times 10^{-19}\,\text{C} \times 90 \times 10^3\,\text{V}$
$= 1.4 \times 10^{-14}\,\text{J}$
Maximum energy of X-ray photon
= electron energy $= 1.4 \times 10^{-14}\,\text{J}$

d Photon energy $E = \dfrac{hc}{\lambda}$

Minimum wavelength λ
$= \dfrac{6.6 \times 10^{-34}\,\text{J s} \times 3.0 \times 10^8\,\text{m s}^{-1}}{1.4 \times 10^{-14}\,\text{J}}$
$= 1.4 \times 10^{-11}\,\text{m}$

2 a See Figure 6.3a.

d The half-value thickness of 2.2 mm causes the emergent intensity I of the X-ray beam to be half of the incident intensity I_0.

$\dfrac{I}{I_0} = e^{-\mu x}$

With $x = 2.2 \times 10^{-3}\,\text{m}$ and $\dfrac{I}{I_0} = 0.5$,
$0.5 = e^{-\mu \times 2.2 \times 10^{-3}}$
$\ln 0.5 = -\mu \times 2.2 \times 10^{-3}$
$\therefore \mu = 315\,\text{m}^{-1}$

4 a Power consumption $P = V \times I$
$= 70 \times 10^3\,\text{V} \times 50 \times 10^{-3}\,\text{A}$
$= 3.5\,\text{kW}$

b Power of X-ray beam = power consumption × efficiency
$= 3.5 \times 10^3\,\text{W} \times \dfrac{0.95}{100}$
$= 33\,\text{W}$

c Area of beam, $A = \pi(\text{diameter}/2)^2$
$= \pi(1.0 \times 10^{-3}/2)^2\,\text{m}^2$
$= 7.85 \times 10^{-7}\,\text{m}^2$

Intensity $I = \dfrac{P}{A}$
$= \dfrac{33}{7.85 \times 10^{-7}} = 4.2 \times 10^7\,\text{W m}^{-2}$

d $\dfrac{I}{I_0} = e^{-\mu x}$
With $\mu = 693\,\text{m}^{-1}$ and $\dfrac{I}{I_0} = 0.125$,
$0.125 = e^{-693x}$
$\ln 0.125 = -693x$
$\therefore x = 3.0\,\text{mm}$

Chapter 7

1 a Activity $A = A_0 e^{-\lambda t}$
Decay constant λ for iodine-131 $= 0.0861\,\text{day}^{-1}$
$= 0.0036\,\text{h}^{-1}$

Initial activity $A_0 = A e^{\lambda t}$
$= 3.7 \times 10^5\,\text{Bq} \times e^{0.0036 \times 36}$
$= 4.2 \times 10^5\,\text{Bq}$

b Decay constant λ for radon-220 $= 0.0127\,\text{s}^{-1}$
Activity $= \lambda N = 7.3\,\text{Bq}$

$\therefore \quad N = \dfrac{7.3}{0.0127} = 575$

Number density $= \dfrac{575}{63\,\text{cm}^3} = 9.1\,\text{cm}^{-3}$

c Decay constant λ for phosphorus-32 $= 0.0475\,\text{day}^{-1}$
$= 0.0020\,\text{h}^{-1}$

Initial activity $A_0 = A e^{\lambda t}$
$= 3.2\,\text{MBq} \times e^{0.0020 \times 120}$
$= 4.07\,\text{MBq}$

2 a Decay constant $\lambda = \dfrac{\ln 2}{t_{\frac{1}{2}}}$
$= \dfrac{0.693}{67}\,\text{h}^{-1}$
$= 0.0103\,\text{h}^{-1} = 2.87 \times 10^{-6}\,\text{s}^{-1}$

b Activity $= \lambda N$ where N is the number of nuclei

$$\therefore N = \frac{6.5 \times 10^8}{2.87 \times 10^{-6}} = 2.26 \times 10^{14}$$

$$\text{Number of mol} = \frac{2.26 \times 10^{14}}{6.0 \times 10^{23}} = 3.77 \times 10^{-10}\,\text{mol}$$

$$\text{Mass} = 3.77 \times 10^{-10}\,\text{mol} \times 99\,\text{g mol}^{-1} = 3.73 \times 10^{-8}\,\text{g}$$

c Activity $A = A_0 e^{-\lambda t}$
$= 6.5 \times 10^8\,\text{Bq} \times e^{-0.0103 \times 72}$
$= 3.09 \times 10^8\,\text{Bq}$

3 a After $(180 - 12) = 168$ hours, the activity is reduced to 0.40 of its maximum.
$A = A_0 e^{-\lambda t}$
$0.40 = e^{-\lambda \times 168}$
giving $\lambda = 5.45 \times 10^{-3}\,\text{h}^{-1}$

$$t_{\frac{1}{2}} = \frac{\ln 2}{\lambda}$$

$$= \frac{0.693}{5.45 \times 10^{-3}} = 127\,\text{h} = 5.3\text{ days}$$

This is the effective half-life.

b $$\frac{1}{T_E} = \frac{1}{T_B} + \frac{1}{T_P}$$

where T_E represents the effective half-life,
T_B represents the biological half-life,
T_P represents the physical half-life.

$$\therefore \frac{1}{T_B} = \frac{1}{5.3} - \frac{1}{14}$$

giving $T_B = 8.5$ days

4 a i) Activity of $5.0\,\text{cm}^3$ of diluted mixture $= 600\,\text{Bq}$

$$\therefore \text{Activity of sample} = \frac{1000 + 10}{5} \times 600\,\text{Bq} = 121\,200\,\text{Bq}$$

$$\therefore \text{Activity of a 5.0 cm}^3\text{ sample} = \frac{5}{10} \times 121\,200\,\text{Bq} = 60\,600\,\text{Bq}$$

ii) Activity of $5.0\,\text{cm}^3$ blood $= 90\,\text{Bq}$
Activity of entire volume of blood = activity of $5.0\,\text{cm}^3$ injected substance $= 60\,600\,\text{Bq}$

$$\therefore \text{Volume of blood} = \frac{60\,600}{90} \times 5\,\text{cm}^3 = 3400\,\text{cm}^3$$

b The effective half-life T_E of iodine-131 is given by

$$\frac{1}{T_E} = \frac{1}{21\text{ days}} + \frac{1}{8\text{ days}}$$

giving $T_E = 5.8$ days. This is long compared with the 20 minutes sampling time, so there is no need to consider the half-life in the calculation.

c Mixing is complete/the radioisotope is spread evenly around the body.

5 a ${}^{23}_{11}\text{Na} + {}^{1}_{0}\text{n} \rightarrow {}^{24}_{11}\text{Na}$

b *Advantage*: limited range in tissue.
Disadvantage: half-life is short so little time for preparation.

6 Minimum activity of technetium sample
$= 0.05 \times$ initial activity of molybdenum
The extracted sample has 90% of the activity of the molybdenum source.
$\therefore$ minimum activity of molybdenum at final extraction

$$= \frac{0.05}{0.9} = 0.0556 \times \text{initial activity}$$

$$\text{Decay constant } \lambda = \frac{\ln 2}{t_{\frac{1}{2}}} = \frac{0.693}{67} = 0.0103\,\text{h}^{-1}$$

$A = A_0 e^{-\lambda t}$

So at final extraction

$0.0556 = e^{-0.0103 \times t}$

giving $t = 280$ hours

$$\therefore \text{number of extractions} = \frac{280}{36} = 7 \text{ (rounded down)}$$

Chapter 8

1 b i) $6.5 \times 10^{-14}\,\text{A} = 6.5 \times 10^{-14}\,\text{C s}^{-1}$

$$\text{This is } \frac{6.5 \times 10^{-14}}{1.6 \times 10^{-19}}\,\text{electrons s}^{-1} = 4.1 \times 10^5\,\text{electrons s}^{-1}$$

ii) $$\frac{4.1 \times 10^5}{4^9} = 1.6\,\text{electrons s}^{-1}$$

iii) $1.6 \times 5 = 8.0\,\text{photons s}^{-1}$
iv) $8.0 \times 10 = 80\,\text{beta particles s}^{-1}$

5 b $$\text{Intensity} = \frac{\text{power}}{\text{area}}$$

Intensity I_1 at screen 1 of area $0.25\,\text{m}^2$ is given by

$$I_1 = \frac{P}{0.25}$$

Intensity I_2 at screen 2 of area $2.5 \times 10^{-4}\,\text{m}^2$ is given by

$$I_2 = \frac{P}{2.5 \times 10^{-4}}$$

As the power at each screen is assumed to be the same,

$I_1 \times 0.25 = I_2 \times 2.5 \times 10^{-4}$

$$\therefore \text{fractional increase in intensity } \frac{I_2}{I_1} = \frac{0.25}{2.5 \times 10^{-4}} = 1000$$

Chapter 9

2 a i) Acoustic impedance for bone,
$Z_1 = 7.78 \times 10^6\,\text{kg m}^{-2}\,\text{s}^{-1}$
Acoustic impedance for soft tissue,
$Z_2 = 1.63 \times 10^6\,\text{kg m}^{-2}\,\text{s}^{-1}$

$$\text{Fraction reflected} = \frac{(Z_2 - Z_1)^2}{(Z_2 + Z_1)^2}$$

$$= \frac{(1.63 \times 10^6 - 7.78 \times 10^6)^2}{(1.63 \times 10^6 + 7.78 \times 10^6)^2}$$

$$= 0.427$$

ii) Fraction that penetrates $= 1 - 0.427 = 0.573$
iii) As ultrasound is going from bone into soft tissue, the fraction reflected is the same as it was in **a** i) when going from soft tissue into bone, i.e. 0.427.
iv) 0.573 of the intensity at the bone–soft tissue boundary is therefore transmitted. So as 0.573 of the original intensity at the front edge of the bone arrives at the back edge of the bone, and 0.573 of this value is transmitted, then $0.573 \times 0.573 = 0.33$ of the original intensity is transmitted through the back edge of the bone.

3 Time taken for ultrasound to travel through the organ and back $= 3.4\,\text{cm} \times 0.050\,\text{ms cm}^{-1} = 0.17\,\text{ms}$
$= 1.7 \times 10^{-4}\,\text{s}$
The speed of ultrasound in the organ is $1200\,\text{m s}^{-1}$,
so $1200\,\text{m s}^{-1} \times 1.7 \times 10^{-4}\,\text{s} = 0.204\,\text{m}$
= twice thickness of organ
So thickness of organ $= 0.10\,\text{m}$

4 a The formula is $\Delta f = \dfrac{2fv\cos\theta}{c}$

where $\Delta f = 0.85 \times 10^3\,\text{Hz}$, $c = 1.5 \times 10^3\,\text{m s}^{-1}$, $f = 5.0 \times 10^6\,\text{Hz}$, $\theta = 45°$.

$$\therefore v = \frac{\Delta f \times c}{2 \times f \times \cos\theta}$$

$$= \frac{0.85 \times 10^3 \times 1.5 \times 10^3}{2 \times 5.0 \times 10^6 \times \cos 45°}\,\text{m s}^{-1}$$

$$= 0.18\,\text{m s}^{-1}$$

5 a Acoustic impedance for air,
$Z_1 = 4.3 \times 10^2\,\text{kg m}^{-2}\,\text{s}^{-1}$
Acoustic impedance for skin,
$Z_2 = 1.7 \times 10^6\,\text{kg m}^{-2}\,\text{s}^{-1}$

$$\text{Fraction reflected} = \frac{(Z_2 - Z_1)^2}{(Z_2 + Z_1)^2}$$

$$= \frac{(1.7 \times 10^6 - 4.3 \times 10^2)^2}{(1.7 \times 10^6 + 4.3 \times 10^2)^2}$$

$$= 0.999$$

$\therefore$ fraction that penetrates air–skin boundary $= 0.001$

b Acoustic impedance for gel,
$Z_1 = 1.6 \times 10^6\,\text{kg m}^{-2}\,\text{s}^{-1}$
Acoustic impedance for skin,
$Z_2 = 1.7 \times 10^6\,\text{kg m}^{-2}\,\text{s}^{-1}$

$$\text{Fraction reflected} = \frac{(Z_2 - Z_1)^2}{(Z_2 + Z_1)^2}$$

$$= \frac{(1.6 \times 10^6 - 1.7 \times 10^6)^2}{(1.6 \times 10^6 + 1.7 \times 10^6)^2}$$

$$= 9.2 \times 10^{-4}$$

$\therefore$ fraction that penetrates gel–skin boundary $= 0.999$

Chapter 10

4 a $\text{Intensity} = \dfrac{\text{power}}{\text{area}}$

$$= \frac{50}{\pi(0.50 \times 10^{-3}/2)^2}\,\text{W m}^{-2}$$

$$= 2.5 \times 10^8\,\text{W m}^{-2}$$

b Delivered energy $= 50\,\text{W} \times \text{time duration} = 0.15\,\text{J}$

$$\therefore \text{duration of pulse} = \frac{0.15}{50} = 3.0 \times 10^{-3}\,\text{s}$$

5 a Refractive index of the fibre, $n_f = 1.52$
Refractive index of the coating, $n_c = 1.40$
$n_f \sin i = n_c \sin r$
When i is the critical angle c, $r = 90°$

$\therefore 1.52 \sin c = 1.40 \sin 90°$

giving $c = 67°$

b Refractive index of air, $n_a = 1.0$
$n_f \sin i = n_a \sin r$
When $i = c$,

$1.52 \sin c = 1.0 \sin 90°$

giving $c = 41°$

Index